Farmer Research Groups

Farmer Research Groups
Institutionalizing participatory agricultural research in Ethiopia

Edited by Dawit Alemu, Yoshiaki Nishikawa, Kiyoshi Shiratori and Taku Seo

Practical Action Publishing Ltd
The Schumacher Centre,
Bourton on Dunsmore, Rugby,
Warwickshire, CV23 9QZ, UK
www.practicalactionpublishing.org

© Dawit Alemu, Yoshiaki Nishikawa, Kiyoshi Shiratori and Taku Seo and the individual contributors, 2016

The right of the editors to be identified as authors of the editorial material and of the contributors for their individual chapters have been asserted under sections 77 and 78 of the Copyright Designs and Patents Act 1988.

All rights reserved. No part of this publication may be reprinted or reproduced or utilized in any form or by any electronic, mechanical, or other means, now known or hereafter invented, including photocopying and recording, or in any information storage or retrieval system, without the written permission of the publishers.

Product or corporate names may be trademarks or registered trademarks, and are used only for identification and explanation without intent to infringe.

A catalogue record for this book is available from the British Library.
A catalogue record for this book has been requested from the Library of Congress.

ISBN 978-1-85339-900-8 Hardback
ISBN 978-1-85339-901-5 Paperback
ISBN 978-1-78044-900-5 Library Ebook
ISBN 978-1-78044-901-2 Ebook

Citation: Alemu, Dawit, Nishikawa, Yoshiaki, Shiratori, Kiyoshi and Seo, Taku (eds) (2016) *Farmer Research Groups: Institutionalizing Participatory Agricultural Research in Ethiopia*, Rugby, UK: Practical Action Publishing, <http://dx.doi.org/10.3362/9781780449005>

Since 1974, Practical Action Publishing has published and disseminated books and information in support of international development work throughout the world. Practical Action Publishing is a trading name of Practical Action Publishing Ltd (Company Reg. No. 1159018), the wholly owned publishing company of Practical Action. Practical Action Publishing trades only in support of its parent charity objectives and any profits are covenanted back to Practical Action (Charity Reg. No. 247257, Group VAT Registration No. 880 9924 76).

The views and opinions in this publication are those of the author and do not represent those of Practical Action Publishing Ltd or its parent charity Practical Action. Reasonable efforts have been made to publish reliable data and information, but the authors and publisher cannot assume responsibility for the validity of all materials or for the consequences of their use.

Cover photo: Members of a Farmer Research Group measure irrigated tomato plants.
Credit: K. Shiratori.
Cover design by Andrew Corbett
Indexed by Liz Fawcett
Printed in the UK

Contents

Tables and figures	vii
1 Introduction: institutionalizing participatory approaches in Ethiopia's agricultural research system *Kiyoshi Shiratori and Dawit Alemu*	1
Part I The need for Farmer Research Groups in Ethiopia	**11**
2 Overview of Farmer Research Group-based participatory research in Ethiopia *Dawit Alemu and Kiyoshi Shiratori*	13
3 FRG-based approach implementation processes: guidelines, training and research *Taku Seo*	25
Part II Experiences of Farmer Research Groups	**43**
4 Engaging farmers in technology evaluation and promotion: Farmer Research Groups on common beans in the Central Rift Valley, Ethiopia *Endeshaw Habte, Kidane Tumsa, Berhanu Amsalu Fenta, and Abiy Tilahun*	45
5 Lowering teff seeding rate using a seed spreader via the participatory approach in South Ethiopia *Fanuel Laekemariam, Gifole Gidago and Wondemeneh Taye*	69
6 Participatory evaluation of selected fish processing and preservation technologies: the case of Lake Tana, Ethiopia *Shewit Gebremedhin, Markos Budusa, Adamu Yimer, Minwyelet Mingist, Dereje Tewabe and Zurihun Nigussie*	89
7 Participatory evaluation of farmer-saved and purified seed for improved agronomic performance: wheat, South eastern Tigray *Alem G/tsadik, Kelali Haftu, Yoshiaki Nishikawa, Ibrahim Fitiwy and Taku Seo*	103
8 Improved dairy production and changing gender roles: experience of smallholder FRGs in Melkassa, Central Rift Valley *Bedru Beshir*	117
9 Farmers' perceived benefits of FRG-based research: the case of selected FRG-based research activities *Shingo Takeda*	133
Part III Institutionalizing Farmer Research Groups in Ethiopia	**145**
10 The participatory approach and FRG: the institutionalization process within the Ethiopian agricultural research system *Dawit Alemu and Kiyoshi Shiratori*	147

11 The challenges of FRG-based research: attitudes, capacity and institutional arrangements 155
Dawit Alemu, Taku Seo, Terutaka Niide, Shingo Takeda, Kiyoshi Shiratori and Yoshiaki Nishikawa

12 Applying the FRG approach in agricultural extension: lessons from the Farmers Research and Extension Group approach 165
Belay Kassa and Dawit Alemu

Part IV Conclusion 183

13 Conclusion: Recommendations for strengthening the responsiveness of agricultural research systems 185
Dawit Alemu, Kiyoshi Shiratori, Taku Seo, and Yoshiaki Nishikawa

Index 201

Tables and figures

Tables

2.1	Comparison of key features of participatory research approaches	21
3.1	Thematic areas and number of research projects supported, 2010–14	34
3.2	Number of research proposals submitted and selected for support, 2010–12)	34
4.1	Important features of the bean varieties used for the participatory evaluation with FRG farmers	48
4.2	Preferred common bean varieties for field testing	53
4.3	Yields of common bean varieties tested at the three sites	57
4.4	Ranking matrix of common bean varieties using qualitative criteria (before field test) in their order of importance as evaluated by farmers	58
4.5	Summary of farmers' assessment of common bean varieties	61
4.6	Summary of the performance and farmers' preference of tested common bean varieties at the three sites	62
4.7	Seed multiplication exercise (with / without the bean management practices trial)	63
4.8	Farmer-based seed production	63
4.9	Challenges observed during the implementation of the participatory variety evaluation process and potential options for better performance	65
5.1	Development and growth characteristics of teff for different seeding rates mixed with a spreader, 2010 and 2011	77
5.2	Grain yield, yield attributing parameters and harvest index of teff for different seeding rates mixed with spreader, 2010 and 2011	78
5.3	Summary of farmers' preferences during crop stand evaluation, 2010 and 2011	80
5.4	Growth and yield attributing parameters and grain yield of teff at lower seed rates, 2012	82
5.5	Partial economic analysis of grain and straw yield, 2012	83
5.6	Farmers' preference evaluation at crop maturation, 2012	83

5.7	Benefit to cost ratio of farmers, preferred teff seeding rates, 2010–12	84
5.8	Farmers, crop stand preference at crop maturation, 2010–12	84
6.1	Sensory evaluation of fish samples processed by selected methods	96
6.2	Microbial load of fish samples with each treatment	97
6.3	Effect on shelf life of microbial load of solar-dried fish	98
7.1	Yields of purified and non-purified seeds of Kubsa, Hawi and Shehan varieties from 2011 to 2013	111
7.2	Grain and biomass yields of Hawi, Kubsa and Shehan varieties, 2013	112
7.3	Interaction effect between seed purification and varieties on grain yield, 2013	112
7.4	Interaction effect between seed purification and varieties on biomass yield, 2013	113
7.5	Evaluation of certified seeds and third-generation, farmer-saved purified and non-purified seeds of Hawi variety	113
8.1	Population of the study kebeles by gender, age and religion	120
8.2	The roles of partners in improved dairy cow introduction and management	122
8.3	Improved dairy farm household composition by age and gender	122
8.4	Average land ownership and land allocation of improved dairy farmers	123
8.5	Types of livestock kept by respondents	123
8.6	Lactation period and average milk yield of local and crossbred cows	124
8.7	Gender role-sharing in livestock management in male-headed households	125
8.8	Gender role-sharing in livestock management in female-headed households	125
8.9	Costs and benefits of keeping crossbreed (Boran x Jersey) dairy cattle	128
8.10	Synthesis of the comparison of local breed and crossbreed dairy cows in the study area	129
9.1	Research targeted for assessment, date of the interview, and number of interviewed farmers, 2014	135
9.2	Contents of the interview and objectives	136

9.3	Examples of reasons for increased yield and land area for cultivation, matched with farmer's perceived benefits	138
12.1	Distribution of respondents by region, zone and *woreda*	168
12.2	Average number of member farmers and proportion of women members in FREGs by region	169
12.3	Distribution of FREGs by region and activity	170
12.4	Member farmers' responses on FREG membership and associated issues	171
12.5	Ratings of member farmers about their involvement in FREG-related decisions	172
12.6	Average number of joint evaluations made on the performance of field experiments (farmers, DAs and researchers)	172
12.7	Perceived benefits of FREG to member farmers	173
12.8	Budget allocated by the RCBP for FREG-related activities and its utilization	176

Figures

1.1	Agricultural research centres in Ethiopia	4
3.1	The framework of the FRG trainings	28
3.2	The support process for FRG research activities	33
4.1	The participatory selection process of bean varieties	53
5.1	Average grain and straw yield of teff at lower seed rates using a seed spreader during 2010–12	83
6.1	The traditional drying method (a) and solar tents (b)	92
7.1	Farmer pouring seeds into salted water	107
7.2	Spoiled seeds and other foreign material floating on the surface of the salt solution	107
7.3	Purified seeds being dried in the shade before planting	107
8.1	A dairy FRG member feeding her Boran x Jersey heifer in open space in Wakie-Mia-Tiyo, May 2014	126
8.2	A dairy FRG member caring for her Boran Jersey cows in a concrete built house in Awash Bishola, May 2014	126
9.1	Farmers' reasons for technology adoption	139
9.2	Farmers' reasons for non-adoption of technology	139
9.3	Challenges encountered in cultivation of subject crop, as perceived by farmers	140
9.4	Type of benefits obtained from the research	141
9.5	Benefits perceived by farmers	141
13.1	Three dimensions of participatory agricultural research	195

CHAPTER 1

Introduction: Institutionalizing participatory approaches in Ethiopia's agricultural research system

Kiyoshi Shiratori and Dawit Alemu

Abstract

In Ethiopia, teaching farmers has been a feature of agricultural research and extension for a long time. Solutions to problems in the form of technologies developed at research stations used to be handed to farmers as passive recipients. The low rate of technology adoption by farmers through the conventional approach led Ethiopian researchers to recognize that research needed to be more demand-driven, through using participatory approaches developed in the 1980s and 1990s. Two technical cooperation projects, FRG I and FRG II, were implemented by Ethiopian Institute of Agricultural Research (EIAR) and Japan International Cooperation Agency (JICA) with the objectives of validating the FRG approach, then scaling up and institutionalizing the approach within the national agricultural research system. This book documents Ethiopian researchers' attempts to develop the FRG approach as part of their normal research practice. The practical experience arising from participatory research documented in the book will be a useful reference to those researchers and research institutions wishing to implement and institutionalize the approach within their own countries and contexts.

Keywords: participatory research, Ethiopia, national agricultural research system, Farmer Research Groups, FRGs, institutionalization

History of participatory research in Ethiopia

In Ethiopia, teaching farmers has been a feature of rural development in agricultural research and extension services for a long time. Solutions to problems faced by farmers were considered to be the responsibility of researchers, and farmers were seen as mere recipients of the technologies and knowledge developed by those researchers at research stations. The agricultural research institutions in Ethiopia, federal and regional research centres and universities with agricultural faculties, have been trying to deliver effective services to farmers in collaboration with the extension system mainly using a linear model, where the research generates technologies and the extension services deliver the developed technologies to the farmers. Technologies developed under such circumstances, however, have often failed to achieve the expected rate of adoption by farmers (Gemechu et al., 2002). This has led to low agricultural productivity, food insecurity and poverty, particularly in rural areas of the country. A small number of researchers concerned with the situation began discussing an alternative approach that might provide better and more effective solutions for farmers.

The importance of more demand-driven research through participatory approaches gained recognition among researchers in Ethiopia throughout the 1980s and 1990s. Although conventional research was still dominant, this shift brought in new, more holistic, farmer needs-based approaches. Farming Systems Research (FSR) was introduced in 1984 with support from the International Development Research Centre (IDRC) of Canada, firstly as a pilot project at two research centres (Bako and Nathareth, now Melkassa) of the then Institute of Agricultural Research (IAR), and later expanded to more centres. It was followed by Client-Oriented Research (COR), piloted in 1998 in the Cool Season Food and Forage Legumes Project and the Africa Highland Initiative. They were implemented within the public research system to enable farmers to participate in research activities (*Agaje et al, 2002; Amanuel et al, 2004*). Non-governmental organizations (NGOs) in collaboration with public research institutions promoted project-based participatory approaches, such as the Farmers' Research Project (FRP), the Institutionalization of Farmer Participatory Research Project, and the Indigenous Soil and Water Conservation Project (ISWC) *(Ejigu et al, 2005; Farm Africa, 2001)*.

The Farmer Research Group (FRG) was used as a component of FSR and more extensively applied in other projects, such as the Joint Vertisol Project and the Participatory Research for Improved Agro-ecosystems (PRIAM) (*Adamo, A.K., 2001; Adugna, 1999; Frew, 1999*). Encouraging results, although limited in scale, were observed in their early activities, particularly in the improved interaction among researchers, farmers and other stakeholders *(Bedru et al, 2009)*. The communication between farmers/pastoralists and researchers in research activities was, however, still constrained by lack of experienced researchers as well as a lack of clear guidelines regarding how to work with FRGs. In most cases, on-farm research activities were limited to the demonstration of new technologies. Regardless, participatory research was generally recognized as an improvement of extension and technology dissemination methods, even though the participatory activities were initially limited in scale and number and were often considered to be outside the official research mandate.

FRG projects

In order to make the agricultural research system more responsive to the needs of farmers and pastoralists, and to recognize them as innovators through improved participatory research. EIAR implemented the first project, in collaboration with Japan International Cooperation Agency (JICA) and Oromia Agricultural Research Institute (OARI), namely Strengthening Technology Development, verification, Transfer and Adoption through

Farmer Research Group (FRG I), which took place between 2004 and 2009, and the second project, in collaboration with JICA, namely Enhancing Development and Dissemination of Agricultural Innovations through Farmer Research Groups (FRG II), between 2010 and 2015. Both projects improved existing methods of implementing demonstration-focused, on-farm activities with FRGs into a research approach, namely the FRG approach, which can function within existing research systems in such a way that any researchers can apply the approach as part of their normal research activities.

The FRG approach is a participatory agricultural research approach in which a group of farmers, development agents and a multidisciplinary research team are jointly involved in the generation, verification and improvement of a technology so as to meet farmers' needs and strengthen their capacity to innovate, thereby improving farm management practices and productivity as well as improving the quality of the research itself. The FRG approach requires researchers to conduct on-farm participatory research with farmer groups so that the technologies generated as research outputs are more appropriate and eventually ensure better technology adoption.

The projects followed two stages. Firstly, FRG I developed the FRG approach, testing and validating it within the existing research framework, with two research centres, the Melkassa Agricultural Research Centre (MARC) of EIAR and the Adami Tulu Agricultural Research Centre (ATARC) of Oromia Agricultural Research Institute (OARI) in the Central Rift Valley. As the second stage, FRG II promoted the approach within the country's entire national agricultural research system (NARS), which consists of EIAR, the regional agricultural research institutes (RARIs) and universities with agricultural faculties. Figure 1.1 shows the location of targeted agricultural research institutions in the country. While the experiences from FRG I and FRG II have shown that the FRG approach can be applied within the existing research system, and that it can lead to functional research-extension and farmer linkages and boost farmers' innovativeness, the approach demands additional researchers' capacity to communicate with farmers, and their institutions to facilitate interdisciplinary implementation of FRG-based research activities.

Objectives of this book

The idea for this book emerged from discussions among the authors about further improving, institutionalizing and scaling up participatory research within the existing formal agricultural research system of Ethiopia. There is much literature on participatory research which discusses the importance of farmers' participation in innovation with actual cases (*Collinson, 2000; Farm Africa, 2001; Reij et al, 2001; Scoones et al, 2009; Wattasinha et al, 2014*). From our own experience, it has been observed that a number of FRG-based research activities, particularly with less experienced researchers, suffer from a lack of capacity to handle on-farm activities aligning scientific research methods with the principles of participation. As the devil is in the detail, the authors of

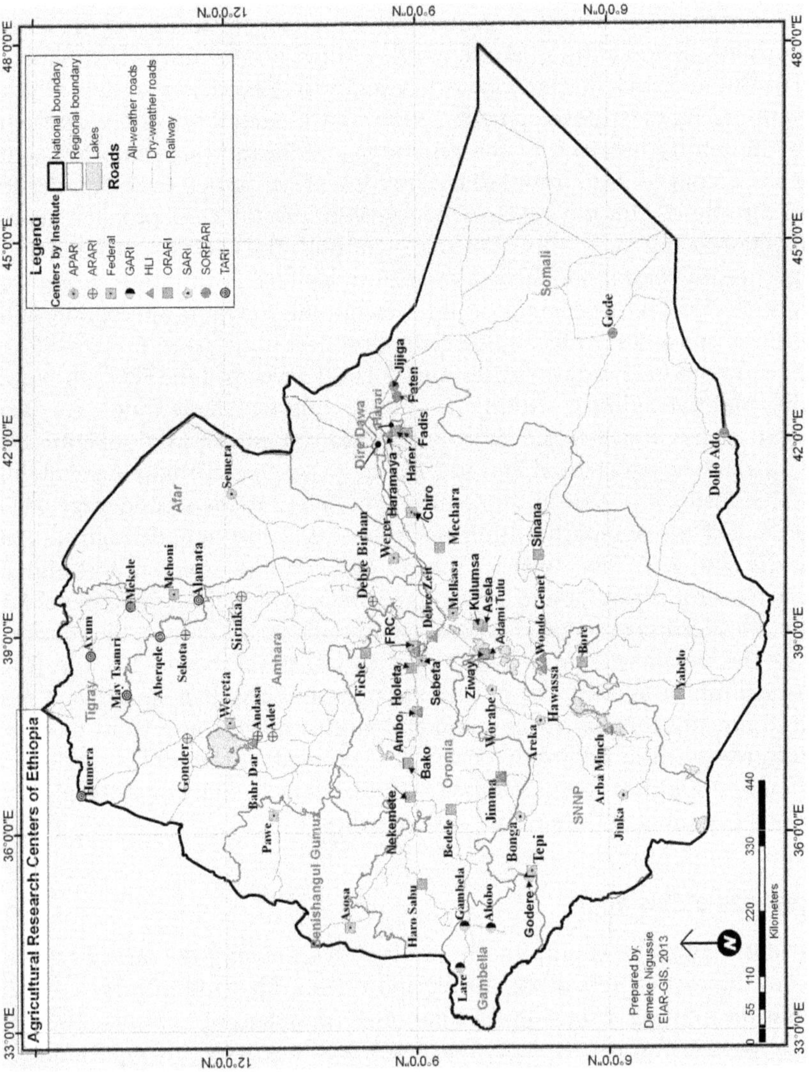

Figure 1.1. Agricultural research centres in Ethiopia
Source: EIAR, 2013

this book felt that researchers needed detailed examples of handling data and processes, combining scientific methods and participatory approaches so as to improve their practice in research.

Available information on participatory research is often very general in this regard and the more recent work focuses on aspects of management, systems and frameworks, tending to lack technical evidence of innovation. Documented examples are also limited in the research areas covered and are often difficult to access by researchers. Most examples in the available literature are based on experiences in donor-funded, project-based, rural development-oriented projects. Such projects with relatively ample funding, usually provide resources and an enabling context for implementation during the project period which often cannot be sustained within the local agricultural research system once the project ends. FRG-based research activities are operated by researchers who face a number of limitations on the ground, including limited and unreliable budgets, limited means of transport, a heavy workload, and a lack of communication facilities, among others. Such limitations make it difficult for researchers to be fully responsible for facilitating multi-stakeholder set-up of innovation systems. Yet, there is widespread interest among development partners and donor agencies, as well as a desire to understand the successes and difficulties of Ethiopian researchers, in implementing the FRG approach.

This book therefore documents the efforts of Ethiopian researchers as they have attempted to develop the FRG approach as part of their normal research practice, within the context of the Ethiopian agricultural research system. At the same time, they have had to minimize the impact of exceptional financial and institutional conditions during the project period.

The FRG approach comprises three interrelated dimensions. The first is that the research outputs must contribute directly to the participating farmers and their communities, helping to solve the problems they face in their specific agro-ecological and the socio-economic context. The second dimension is that the process should follow normal research protocols, so that the outputs can be scientifically validated and published as with any other research activities carried out at research stations. The third dimension is to develop the farmers' own capacity to innovate, owing to the knowledge and skill gained during the period of the FRG activities. This book provides practical evidence of how these three dimensions have been developed in Ethiopia, and demonstrates that participatory approaches can be deployed within the existing research system. The book also provides useful information about institutionalizing participatory approaches within that formal agricultural research system.

Summary of the contents

This book brings together the experiences and reflections of those who have been involved in the two projects applying the FRG approach, and who have promoted it nationally within the Ethiopian agricultural research system.

The book is divided into four parts and each part consists of several chapters. Part I, with three chapters, deals with the background of the participatory approach in Ethiopian agricultural research, explaining its evolution and present status. This introductory chapter gives a brief history of participatory research in Ethiopia and the interventions that have taken place through FRG I and FRG II. Chapter 2 provides a brief description of the FRG approach with its principles and processes, and the characteristics of the approach in comparison with other participatory approaches. Chapter 3 deals with the interventions made by FRG I and FRG II and describes the detailed steps of promoting and institutionalizing the FRG approach, with practical cases of successes and failures. It first describes the steps taken and strategies used for developing the FRG approach guidelines. The author then shows how the FRG training programme has been arranged and implemented, including institutional arrangements, with six training hubs conducting training for trainers and curriculum development. It is followed by an overview of the project's support for FRG-based researchers and identifies specific issues and measures taken to address them.

Part II, with six chapters, presents practical case studies of FRG-based research activities and monitoring results. Four chapters present research in different disciplines in various parts of Ethiopia, with in-depth descriptions of the process and relevant scientific information bean variety evaluation by Melkassa Agricultural Research Centre (Chapter 4); the seeding rate of teff, Ethiopia's staple crop, by Wolaita Sodo University (Chapter 5); fish processing by Bahir Dar University (Chapter 6); and farmer-saved seed by Mekelle University (Chapter 7). They also share actual experiences on the difficulties of the various stages of participatory research. Since participatory agricultural research is a form of scientific research, it is necessary for the researchers to follow the conventions of scientific writing to some extent. It is hoped that readers understand that this can be difficult to achieve in many circumstances, and that their own contributions could be integrated to boost further improvement of participatory research. These case studies are followed by two chapters of monitoring results to share the merits and limitations of the FRG approach. Chapter 8 reports on an evaluation of a dairy research activity, implemented by MARC. Chapter 9 presents the results of a survey of 13 FRG-based research activities on the benefits farmers have perceived through their participation. The majority of farmers reported benefits and one-third had adopted the technologies targeted by the research. These chapters show that FRG-based research can remain as a part of the formal research process, through presenting scientific procedures and outputs with detailed explanations of participation arrangements.

Part III, with three chapters, deals with the institutionalization of the FRG approach in the public agricultural research system and includes a detailed analysis of the situation at the time of writing. Chapter 10 describes the attitude and capacity of researchers and the institutionalization process, discussing the challenges associated with researchers' perceptions of farmer participation and the skills required to deliver participatory research. Chapter 11 presents the strategies and measures being taken by research institutes under federal and regional governments to promote participatory research, along with the challenges faced in the process of institutionalizing the FRG approach. Chapter 12 introduces the adoption of the FRG approach by different agricultural research and extension projects, such as the Rural Capacity Building Project (RCBP), the East African Agricultural Productivity Project and the Pastoralists Community Development Project of the World Bank in Ethiopia. It provides information on how the approach has been applied, with detailed analysis of the Farmer Research and Extension Group (FREG) of RCBP. Since FREG is a farmer participatory approach derived from FRG but more focused on demonstration and extension, there has been some discussion and confusion about what FRG and FREG are. This part provides some insight on appropriate approaches in agricultural innovation.

Part IV, with a concluding Chapter 13, summarizes the potential and the challenges of promoting and institutionalizing the FRG approach in the Ethiopian agricultural research system by covering the research-extension linkages, gender, researchers' behavioural changes, and local culture and social norms. It concludes with recommendations to strengthen the responsiveness of agricultural research systems, with specific reference to Ethiopia and Africa at large.

Significance for development studies and participatory research

'Participation' is a frequently used word within the development community, although enthusiasm for it may be shifting to a minor key in today's development agenda, compared to the late 1990s and early 2000s. This is not because the participatory approach has already been mainstreamed, or that it has been proved to be ineffective, or that it has marginal importance in the current agenda of agricultural research. Rather, it is because it has lacked favour with the prevalence of output-oriented evaluation systems, as the participatory approach and participatory research in particular is a comparatively cumbersome process, dealing with a wide variety of issues in practical implementation. Another reason is simply that it has not been tried rigorously enough because its methodology is not as straightforward as conventional research.

This book presents practical cases of the participatory approach in agricultural research and its institutionalization in the standard research framework, rather than theories of what is supposed to be done, or featuring researchers in only marginal roles in the participatory process. It not only discusses technology generation, but also the processes which lead to innovation, and how researchers and institutions facilitate them. The communication and linkages among and between stakeholders is also a main component of the book.

It is hoped that the practical experiences of the participatory research approach examined in the book, its achievements and challenges, will be a source of encouragement to researchers to try out the participatory approach, and to senior management and decision-makers of research institutions to institutionalize the approach, bring their researchers into the farmers' fields, interact with farmers in research activities, generate technologies and be catalysts for innovation.

About the authors

Kiyoshi Shiratori is a consultant specializing in rural development at Kaihastu Management Consulting. He is a visiting professor at the Graduate School of Asian and Africa Area Studies, Kyoto University. He was a Chief Advisor in the FRG I and FRG II projects in Ethiopia.

Dr Dawit Alemu is Director of the Agricultural Economics, Extension and Gender Research Directorate of the Ethiopian Institute for Agricultural Research. He has been associated with EIAR since 1999 as a Senior Researcher and Coordinator. His research focuses on agricultural marketing with an emphasis on agricultural inputs.

References

Adamo, A. K. (2001) Participatory Agricultural Research Processes in Eastern and Central Ethiopia: Using Farmers' Social Networks as Entry Points, Occasional Publication Series No. 33, Kampala: Centro Internacional de Agriculture Tropical (CIAT) and Ethiopian Agricultural Research Organization. Cali, Colombia: CIAT Occasional Publications Series No. 33. CIAT.

Adugna, W. and Tesfaye, A. (1999) Participatory Research at Nazreth, Ethiopia. In Farley C. Ed. Proceedings of a synthesis workshop, Nazreth, Ethiopia, 17-21 August 1998. Network on Bean Research in Africa, Workshop Series, No. 38. CIAT, Kampala, Uganda.

Agaje T., Chilot Y., and Taye B. (2002) 'Experiences of participatory research in the National Agricultural Research System of Ethiopia', In Gemechu K., Yohannes G., Kiflu B., Chilot Y. And Asgelil D. (eds.), *Towards Farmers' Participatory Research: Attempts and Achievements in the Central Highland of Ethiopia. Proceedings of Client-Oriented Research Evaluation Workshop, 16-18 October 2001*, Holetta Agricultural Research Centre, Holetta, Ethiopia.

Amanuel A., Teshome D., Yohannes G. and Waters-Bayer, A. (eds.) (2004) *Promoting of Farmer Innovations and Experimentations in Ethiopia (PROFIEET). National Workshop Proceedings, August 25-27, 2003*, Ethiopian Management Institute, Debrezeit, Ethiopia.

Bedru B., Berhanu S., Endeshaw, H. Matsumoto, M. Niioka, K. Shiratori, Teha M. and Wole K. (2009) *Guideline to Participatory Agricultural Research through Farmer Research Group (FRG) for Agricultural Researchers*. FRG Project, EIAR, OARI and JICA. Addis Ababa, Ethiopia.

Collinson, M.P. (2000) *A history of farming systems research*, Wallingford: FAO and CABI International,.

Ejigu, J. and Waters-Bayer, A. (2005) *Unlocking Farmers' Potential: Institutionalising Farmer Participatory Research and Extension in Southern Ethiopia*, Project Experience Series, Farm Africa, London.

FARM-Africa (Ethiopia) (2001) *Framer Participatory Research in Southern Ethiopia: the Experience of Farmers' Research Project*, Farm Africa, London.

Frew, M. (1999) 'Participatory Research for Improved Agroecosystem Management: a Community-based Approach in Eastern Ethiopia', Alemaya Wereda. In Farley C. (ed.), *Proceedings of a Synthesis Workshop, Nazreth, Ethiopia*, 17-21 August 1998. Network on Bean Research in Africa, Workshop Series, No. 38. CIAT, Kampala, Uganda.

Gemechu, K., Yohannes G., Kiflu B., Chilot Y. and Asgelil D. (eds.) (2002) *Towards Farmers' Participatory Research: Attempts and Achievements in the Central Highland of Ethiopia. Proceedings of Client-Oriented Research Evaluation Workshop, 16-18 October 2001*, Holetta Agricultural Research Centre, Ethiopia.

Leeuwis, C. and van den Ban, A. (2004) *Communication for rural innovation: rethinking agricultural extension*, third edition. Oxford, Blackwell.

Reij C. and Waters-Bayer, A. (2001) *Farmer innovation in Africa: a source of inspiration for agricultural development*, London, Earthscan.

Scoones, I. and Thompson, J. (2009) *Farmer First Revisited: Innovation for Agricultural Research and Development*, Rugby: Practical Action.

Warren, D. M. and Cashman, K. (1988) *Indigenous knowledge for sustainable agriculture and rural development*. International Institute for Environment and Development, Sustainable Agriculture Programme. London.

Wettasinha C, Waters-Bayer A, van Veldhuizen L, Quiroga G and Swaans K. (2014) *Study on impacts of farmer-led research supported by civil society organizations*. Penang, Malaysia: CGIAR Research Program on Aquatic Agricultural Systems.

PART I
The need for Farmer Research Groups in Ethiopia

CHAPTER 2
Overview of Farmer Research Group-based participatory research in Ethiopia

Dawit Alemu and Kiyoshi Shiratori

Abstract

The Farmer Research Group (FRG) approach is one of the participatory research methods adopted in two projects promoted within the national agricultural research system of Ethiopia (NARS). The first project (2004–09) demonstrated evidence of the advantages of this approach for wider adoption and institutionalization. The second project (2010–15) set about the institutionalization and scaling up of the FRG approach along with capacity building for all actors in the NARS. FRG is an approach whereby a multidisciplinary research team, development agents and groups of farmers jointly conduct research on selected topics, based on farmers' needs, in farmers' fields. Since 2006, the application of the approach has been expanded to all members of the NARS. The approach follows logical steps: 1) establishing a multidisciplinary research team; 2) assessing farmers' needs; 3) identifying technological options; 4) forming the farmer group and identifying relevant stakeholders; 5) joint implementing, monitoring and evaluating; and 6) sharing experience.

Keywords: participatory research, Ethiopia, Farmer Research Groups, FRGs, institutionalization, national agricultural research system (NARS)

Introduction

Following the recognition of the participatory research approach in the 1970s as a mechanism to ensure the proper consideration of farmers' biophysical and socio-economic constraints and the development of technologies that better suit farmers' circumstances, various participatory approaches have been considered in the agricultural research system in Ethiopia. The first was Farming Systems Research (FSR) initiated in 1976 by the then Institute of Agricultural Research through a project that attempted to consider farmers' perceptions and the process of extension activities, and involved dynamic elements, including testing of recommended technology in the farming environment. In subsequent years, the approach was institutionalized with the establishment of the Department of Agricultural Economics and Farming

Systems Research where researchers with social science and agronomy expertise were assigned. Subsequently, biophysical research began in participatory plant breeding and participatory variety selection that ensured farmers' involvement in the research process.

In Ethiopia, along with the promotion of participatory research, there is a parallel endeavour to strengthen the research-extension process. This has been instrumental to further strengthening the role of participatory research. In the second half of the 1990s, an innovative strategy was adopted by the Ethiopian Institute of Agricultural Research (EIAR), intended to make agricultural research and extension activities responsive and relevant to smallholder farmers by involving them in the selection of research and extension priorities and in research planning and implementation through the establishment of Farmer Research Groups (FRGs) (FDRE, 1999). FRGs are voluntary groups that farmers form to experiment with research and extension activities in their own fields. The formation of appropriate groups is based on production constraints as identified and prioritized by the farmers themselves.

The formal promotion of the FRG approach within the national agricultural research system (NARS) was established by a project entitled Strengthening Technology Development, Verification, Transfer and Adoption through Farmers Research Group (FRG I), – implemented jointly by EIAR, Oromia Agricultural Research Institute (OARI) and Japan International Cooperation Agency (JICA) from 2004 to 2009 in the Oromia Region. Given the success of this project and the lessons learned, a second phase – Project for Enhancing Development and Dissemination of Agricultural Innovations through Farmer Research Groups (FRG II) – was implemented from 2010 to 2015 with the revised objectives of institutionalization and scaling up of the approach along with the required capacity building for all actors of the NARS. This chapter provides an overview of the principles and processes of the FRG approach, a comparison with other participatory approaches, and the main framework of the two project phases.

Principles of the FRG approach

The FRG approach is a participatory research approach in which a group of farmers, development agents (DAs) and a multidisciplinary research team jointly participate in agricultural technology generation, verification and improvement so as to meet farmers' needs and improve farmers' production and management practices as identified by themselves. Accordingly, there are four elements essential to the participation of respective actors in the group: collective action, multidisciplinary team formation, information and knowledge sharing, and cost sharing. Based on these elements, the principles of the approach are:

Principles

Multidisciplinary research team formation. The research team with expertise in a variety of disciplines is key, as it allows consideration of the range of complexities for farm household technological innovation. It also helps in the prioritization of the issues involved and the type of engagement needed once they are identified.

Equal partnership. The principle of equal partnership among all involved is crucial to the approach.

Collective action. It is essential that farmers participate as a group. Their collective activities play important roles in: 1) technology development and improvement, 2) technology dissemination, 3) marketing, and 4) promotion of cooperatives.

Building capacity for innovation. FRG activities are geared towards developing farmers' capacity to innovate and DAs' capacity to facilitate the innovation, not only in the area of direct FRG activity but also improving livelihoods with income-generation innovations.

Members' diversity. The fair representation of different categories of farmers based on gender, resource ownership, and age ensures different perspectives are included and there is potential for innovation from different types of farmers.

Information and technology sharing. Proper documentation of every experience and the technologies adapted and/or generated is vital for effective dissemination as these are invaluable resources for other farmers, DAs and researchers.

Cost sharing. Cost sharing is promoted mainly to ensure collective ownership among relevant stakeholders, including farmers, which is vital to ensure the sustainability of the group's activities.

Process of the FRG approach

Planning FRG activity

Once an existing need is recognized, based on observation of current or potential practices or farmers' demand, it can be used as an entry point to initiate FRG research activities with a newly established group. Accordingly, the entry point can be based on the availability of potential technologies, the expressed needs of farmers, needs arising from farmers' own practices, and/or adding value to an ongoing development process.

Establishing a multidisciplinary research team

In general, the FRG research topic and the issues to be addressed determine the composition of the research team. Agriculture is a complex sector, so issues and challenges not addressed by the initial team members can emerge during the implementation process. In such cases, there needs to be an option to revisit the composition of the team to ensure researchers with relevant disciplines are involved.

Assessing farmers' needs

After the initial constraints and challenges have been expressed, the multi-disciplinary research team needs to work with farmers, DAs and other stakeholders to assess each of their needs, interests, and what resources are available, and to identify the potential and feasible areas (technological, process, innovation, linkages etc.) of FRG activities.

Identifying technological options

The possible technological options are often determined based on the following:

Technical feasibility. This is the practical possibility of applying a proposed solution. Can this activity be carried out within the capacity, skill and knowledge of FRG members at the conclusion of the FRG research?

Farming system compatibility. This is about matching any technical options identified with the farming system and in terms of production systems, resource requirement (labour availability, time allocation and other inputs), and considering the whole value chain of a commodity.

Economic feasibility. This deals with the affordability, cost-effectiveness and potential profitability of a proposed research activity to the farmers.

Social feasibility. Is the proposed research suitable or acceptable within the existing social arrangements (e.g. division of labour) of the community?

Expected risks. This means considering the associated risks of the trial, for example, the effect of any chemicals used, or any effects of introduced varieties/breeds on the local ones.

Research capacity. This concerns making an assessment regarding whether the researchers' capacity and the available research facilities are adequate to carry out the particular trials being considered.

Once the options are known, each of them needs an initial assessment. It is important for researchers to have clear ideas about each option's potential advantages, disadvantages and impacts on farming. It is recommended that the possible options are summarized in a table for presenting to DAs and farmers, to help with prioritization and selection.

Forming the FRG and identifying relevant stakeholders

This involves considering who will be the members of the group, its size and how it will be governed. Before the selection of individual farmers, a suitable community within the target area must be selected where the identified need is based on wider relevance, accessibility and commitment. Then, target farmers can be selected in collaboration with the DA, *kebele* (district/village) leaders and/or by holding a community meeting. FRG research is not aimed at community development directly, but is targeted at generating agricultural technologies which will in themselves be key to community development. Experience has indicated that consideration of the following criteria helps to select farmers, in line with the key principles of the FRG approach:

Composition of group members. Consider gender, geographical spread and relative wealth.

Interest/initiative. Select farmers who are ready to try out new ideas. DAs can be key to finding out this information, as well as those who participate at the village meeting.

Willingness to contribute. Select farmers who are ready to share information, realizing that to do so will eventually benefit themselves and the wider community.

Community consensus. It is important that a consensus is reached within the community regarding the selection of participating farmers. This leads to better understanding within the community as well as greater awareness by the farmers themselves of their responsibilities to the community.

The group size is generally recommended to be between 15 and 20 households. However, depending on the subject addressed, the settlement pattern of households and also the interest level within the target community, the group size may be bigger or smaller than this. In general, fewer than 10 households for livestock research topics and more than 30 for weed management and marketing subjects are both practical (Bedru et al, 2009).

Once the group members are selected, group governance can be discussed and agreed. For example, there may be a need for a group chairperson and a secretary. There are two possible ways of selecting for these positions. The first is based on a democratic election and the second is by appointment, based on information and consultation with DAs and *kebele* administrators. Both have their advantages and disadvantages, so it will be important to consult and weigh both options before deciding. The chairperson is responsible for overall coordination between the group members and for liaising between the group and the researchers as well as organizing meetings. The secretary is responsible for record keeping and generally assisting the chairperson to ensure the smooth running of the group.

Identification of stakeholders

In any rural context, there are possible public and private stakeholders that can support the activities of FRG groups or that can benefit if they are aware of or involved in the research. Possible stakeholders include: agricultural offices, farmer training centres, development programmes, traders operating in the area, input suppliers, farmers' cooperatives, and local officials. In order to involve stakeholders, it is important to create awareness about the intent and objectives of the FRG activity.

Joint action planning

Action planning must take place in the presence of all the stakeholders: FRG member farmers, research team members, DAs and other stakeholders. The process must involve listing all the FRG activities that will achieve the agreed objectives, identifying a schedule when each of the activities should be carried out, assigning a responsible person for each activity, and finally setting a timetable for monitoring, evaluation and support.

Joint implementation of field activities

The implementation of FRG field activity follows these key steps:

1. Orientation. The research team provides background information about the activity to promote understanding of the FRG approach among participating farmers and DAs. Farmers also need to be made aware of the necessary procedures to be followed before actual field implementation can begin.

2. Trialling. This involves preparing all the required inputs and the physical placement of trials in the field as per the agreed action plan. As the FRG approach is as much about building farmers' capacity to innovate as it is about the actual trial results, each step of the implementation process has to be accompanied by hands-on training, discussion and consultation between farmers, researchers, DAs and other participating stakeholders. Agriculture is a complex venture, so care has to be taken at all stages of the implementation process to avoid and minimize any failures.

3. Field performance observation, data collection and documentation. These activities are performed by farmers, DAs and researchers. The farmers are expected to learn scientific methods of observation in the field trial. Similarly, DAs and researchers collect required data for the proper documentation of the progress of the trial. For ease of data collection, forms can be prepared ready to fill in. It is the responsibility of the research team to document the progress made and the performance of the trials based on analysis of the collected data.

Monitoring, evaluation and support

Involves undertaking regular meetings, supervision visits, and joint monitoring and evaluation (M&E) visits (by the research team, extension workers, and members of the FRG), to assess progress and to address any emerging issues that might affect achieving the agreed targets. Robust M&E often helps in adjusting FRG activities before larger problems emerge.

Sharing experiences

Experience-sharing activities are often scheduled in the planning phase of any FRG activity. The experience sharing can involve not only members of an FRG but also members across different FRGs and other stakeholders involved. Successful ways to share experiences include field open days, exchange visits, and experience-sharing training programmes.

A field open day is an important way to share experience with wider stakeholders. Farmers open their trial fields for a day, inviting non-FRG member farmers within and outside the community and those working on extension programmes so they can visit and learn. Exchange visits, on the other hand, are where farmers visit each others' trials to share and learn from each other, and can discuss together their progress, issues and lessons learned.

The FRG approach compared to other participatory approaches

Though there are commonalities across participatory research approaches, there are key features that differentiate them. The main approaches considered here are FRG, Farming Systems Research (FSR), Farmer Research and Extension Groups (FREG) and Participatory Technology Development (PTD).

Differences between approaches are often determined based on: 1) the level of participation of researchers, farmers and other stakeholders, 2) the type of tools applied, 3) whether the process promotes group action, 4) the level of consideration of the interest of target farmers, and 5) the expected outputs (Bedru et al, 2009; Chimdo, 2008; Conroy and Sutherland, 2004). Table 2.1 summarizes the key features of the different participatory research approaches.

The level of farmer participation can be passive, active or interactive, depending on the type of approach promoted. Passive participation is when farmers or target communities are simply recipients of messages, assistance and services. Active participation is when farmers or target communities are being consulted, and they provide information on constraints, needs and even possible solutions. Though farmers or communities have choices, final solutions are offered by researchers. Interactive participation is when farmers or communities, either among themselves or jointly with researchers, interact in knowledge exchange, solution finding, and decision taking, in implementation and in monitoring and evaluation. The level of 'ownership' experienced is highest in interactive participation, followed by active participation, and a limited sense of ownership in the case of passive participation. Accordingly, among the different participatory research approaches, FRG, FREG and PTD involve more interactive participation of farmers.

The framework of FRG projects

This section presents the key features of the framework of implementation of two projects that have promoted the FRG approach in Ethiopia since 2004, specifically, the institutional and resource arrangements of FRG I (2004–09) and FRG II (2010–15).

The FRG projects were implemented to promote wider technology adoption by addressing some identified challenges, namely: the lack of appropriate farming technologies, the poor linkage between the research and extension systems, the dominance of non-participatory on-station research, and the limited benefits of conventional research to resource-poor farmers. FRG I was implemented by the Melkassa Agricultural Research Centre of EIAR and the Adami Tulu Agricultural Research Centre of OARI, with the support of JICA. It targeted the East Shewa Zone and part of the Arsi and West Arsi Zones of the Oromia Region. The main objective of the project was to demonstrate evidence for wider scaling up of the approach through diverse activities, including developing the FRG research approach, based on the practical experiences of participatory research through farmer groups with the close collaboration of DAs; promoting client-oriented research topics to develop appropriate technologies; and developing the capacity of farmers, DAs and researchers to generate, modify and disseminate technology using the FRG approach.

FRG II was built on the successes of FRG I and its objectives were to scale up and institutionalize the FRG approach in the national agricultural research system; to support the development of appropriate technologies in priority

Table 2.1 Comparison of key features of participatory research approaches

Feature	Farming Systems Research (FSR)[1]	Participatory Technology Development (PTD)[2]	Farmer' Research Group (FRG)[3]	Farmer Research and Extension Group (FREG)[4]
When started	Developed in the second half of the 1970s	Began in the early 1980s	Began in the late 1990s	2008
Definition /objective	Directly involves the ultimate beneficiaries in on-farm research Develops socio-economic methods to consider the interactions between on-farm and off-farm resource management Recognizes the value of ITK.	Joint experimentation and research by farmers and development agents in discovering ways of improving farmers' livelihoods	A multidisciplinary research team, development agents and groups of farmers jointly conduct research on selected topics based on farmers' needs on farmers' fields	Farmer groups promoted from FRG for further scaling up of technology Member farmers of FREG act as experts who teach other farmers
Tools used	Diagnostic surveys Holistic approach	Use of farmer-innovators Awareness creation on more systematic forms of experimentation and associated capacity building Facilitation of the generation of insights and options within the community	Farmers' group formation Focus group discussions Preference ranking	Farmers' group Training Demonstration Field day
Researchers' involvement	Direct involvement	Limited researcher involvement	Facilitation role Interactive participation	Advisers, resource people
Farmers' level of participation	Passive	Interactive	Interactive	Interactive
Extent of group action	Community-based	Innovator, individual farmer-based	Group-based	Group-based
Target farmers	Interested individual farmers	Interested individual farmers	Only interested and organized members of the FRGs	Interested organized members of FREGs

(continued)

Table 2.1 Key features of participatory research (continued)

Feature	Farming Systems Research (FSR)[1]	Participatory Technology Development (PTD)[2]	Farmer' Research Group (FRG)[3]	Farmer Research and Extension Group (FREG)[4]
Expected outputs	Better understanding of farmers' circumstances by researchers Improved availability of technologies and improved agricultural practices that address farmers' problems	Innovation and adoption of new ways of managing agricultural and natural resources Farmers' lead innovation based on their own priorities	Adapted appropriate technologies Improved adoption of available technologies Improved capacity of farmers to innovate further Enhanced group action Better understanding of farmers' circumstances by farmers Better research targeting	Adoption of technology Technology scale-up Farmer-to-farmer dissemination

Source: 1 FAO. 1994; 2 Conroy and Sutherland, 2004; 3 Bedru et al., 2009; 4 Chimdo, 2008 research topics; and to support and build the capacity of researchers in the development and use of extension materials. The project was centrally managed by project office staff based at EIAR headquarters. Participating research organizations were research centres from both federal and regional institutes, and from higher learning institutes with agricultural faculties.

Conclusion

Participatory research through FRGs started in Ethiopia in the late 1990s. It was then developed into a farmer group-based research approach and promoted in the national agricultural research system between 2004 and 2015 through two projects, FRG I and FRG II. The FRG approach is characterized by group/collective action, multidisciplinary teams and on-farm trials. It focuses on farmers' needs and its outputs must be technically, economically and socially feasible for farmers. The procedures and stages of the FRG approach are jointly implemented with clear responsibilities for researchers, farmers, DAs and other stakeholders in planning, data collection, monitoring, analysis and information sharing. The FRG approach promotes more interactive participation of farmers in its research activities and their ownership of the process and its outputs.

About the authors

Dawit Alemu is Director of the Agricultural Economics, Extension and Gender Research Directorate of the Ethiopian Institute for Agricultural Research. He has been associated with EIAR since 1999 as a Senior Researcher and Coordinator. His research focuses on agricultural marketing with an emphasis on agricultural inputs.

Koyoshi Shiratori is a consultant specializing in rural development at Kaihastu Management Consulting. He is a visiting professor at the Graduate School of Asian and Africa Area Studies, Kyoto University. He was a chief advisor in the FRGI and FRGII projects in Ethiopia.

References

Bedru, B., Berhanu, S., Endeshaw, H., Matsumoto, I. Niioka, M., Shiratori, K., Teha, M. and Wole, K. (eds) (2009) *FRG Guideline For Agricultural Researchers*, Addis Ababa, Ethiopia.

Chimdo, A. (2008) *A Practical Guide to Farmer Research Groups*. Rural Capacity Building Project, Ministry of Agriculture and Rural Development, The Federal Democratic Republic of Addis Ababa, Ethiopia.

Conroy, C. and Sutherland, A. (2004) *Participatory technology development with resource-poor farmers: maximizing impact through the use of recommendation domains*. Network Paper No. 133, The Overseas Development Institute, London, UK.

FAO (1994) *Farming Systems Development: A participatory approach to helping small-scale farmers*. FAO-Sweden Farming Systems Programme in Eastern and Southern Africa. Food and Agriculture Organization of The United Nations. Rome.

The Federal Democratic Republic of Ethiopia (FDRE) (1999) *Agricultural Research and Training Project: Ethiopian Research-Extension-Farmer Linkage Strategy*, Vol. 1, Addis Ababa.

CHAPTER 3
FRG-based research implementation processes: guidelines, training and research

Taku Seo

Abstract

This chapter describes the steps followed by two projects in FRG guideline development, and the provision of training and support for FRG research activities. The FRG guidelines were developed during FRG I and widely applied by research centres and universities. Along with the development of the guidelines, a training program on the FRG approach was organized for researchers from all members of the national agricultural research system. In total, the project has trained more than 600 researchers. During this process, the key challenges in the implementation of the FRG approach to training and research were: staff turnover, the limited research capacity of some researchers, and the limited communication skills of some researchers to share experience with others. It will be important to address these challenges if the approach is to be implemented sustainably.

Keywords: Ethiopia, Farmer Research Group, FRG approach, guidelines, research, training, process of activities

Introduction

The Farmer Research Group projects in Ethiopia were implemented in two phases: Strengthening Technology Development, Verification, Transfer and Adoption through Farmers Research Groups (FRG I) between 2004 and 2009, and the Project for Enhancing Development and Dissemination of Agricultural Innovations through Farmer Research Groups (FRG II), between 2010 and 2015. They both had targeted implementation strategies which were: developing FRG approach guidelines; providing FRG training for researchers within Ethiopia's national agricultural research system (NARS); and supporting FRG-based research activities in selected topics, including the development of extension materials based on research outputs.

These are of course interrelated and complementary to each other. This chapter provides details of the steps and implementation strategies followed during the development of FRG guidelines and the provision of training and support for FRG-based research activities. The chapter is organized into four sections. The first section deals with the approach taken in the development of the guidelines along with the strategy adopted to promote their wider application. The second section presents the approach followed in carrying out the FRG training programmes during the two projects. The third section details the provision of FRG-based research support. The final part outlines the key lessons learned and concluding remarks.

Developing guidelines for the FRG approach

The need for FRG research guidelines emanated from the recognition that enhancing researchers' capacity to apply scientific research methods within a participatory context was key to the success of the approach. The pilot phase of FRG I therefore prioritized the development of such guidelines in parallel with providing training for researchers.

The FRG guidelines were developed by the researchers at Melkassa Agricultural Research Centre and Adami Tulu Agricultural Research Centre, supported by the project. The guidelines were organized into a summary booklet. Through the process of guideline development, the researchers adopted a trial and error approach, with the goal of developing a guide which any researcher could put into practice and be sure of achieving a reasonable standard of participatory research.

When the FRG I project started in 2004, FRG-based activities were already being carried out by researchers. Thus, the project started by improving the processes of existing FRG activity. For example, most FRG-based activities conducted participatory rural appraisal (PRA) in their planning stage, using varied methodologies and degrees of success. After the project had gathered on-the-ground experience, it was recommended that a cropping calendar, resource mapping, a daily calendar, and a gender analysis methods should be standard tools to be used by all PRA research teams. This was made relatively straightforward for inexperienced researchers to initiate contact with farmers. Joint planning with local agricultural offices was also proposed, as well as extending membership to wives of male participants (the opportunity was originally given mainly to female household heads), and a useful set of template forms was produced for researchers to use to deal with commonly experienced challenges.

In terms of facilitating multidisciplinary teams of researchers, after much trial and error, various elements were identified as key for smooth implementation and achievement of tangible research outputs and incorporated into the guidelines: regular meetings (weekly within the team and every one to three months among research teams); careful record keeping of field activities by the researchers; and regular progress review meetings on research activities by farmers and extension staff.

Utilization and improvement of the guidelines during FRG II

FRG guidelines were distributed to all the FRG training participants, all the agricultural research centres and universities with a faculty of agriculture, to be used as a training material. Initially, FRG II was expected to produce different FRG guidelines by region and research topic. This was in response to demand from the training participants for regionally based or topic-based guidelines. However, near the end of FRG II, those researchers who participated in the project as core team members (those who were giving FRG training and those who were practising FRG research activities) concluded that there was no need to develop either regionally based or topic-based guidelines. It was felt that the essence of the approach is the same in any circumstance, and that it is impossible to specify what to do in each location and topic in a general set of guidelines. Rather, they suggested the importance of the guidelines being deliberately general, so that users can contextualize the guidelines as necessary, while being confident in the essence of the approach.

Adoption of the guidelines

In 2014, towards the end of FRGII, the Pastoral Community Development Project Phase III began. This is a five-year World Bank-supported project intended to improve the livelihoods of pastoral communities in Ethiopia. One of the project components was 'Promotion of adaptive research and innovative practices', which the FRG approach fits within. In this regard the 'Guideline to Participatory Agricultural Research through Farmer Research Groups (FRGs) for Agricultural Researchers' was adapted and Pastoralist Agro-Pastoralist Research Group (PAPRG) guidelines were developed to be used for the training and implementation of PAPRG-based research activities in the target areas.

Experienced trainers in the FRG approach provided the introductory training of the PAPRG research approach. Although specific PAPRG research guidelines have been developed, they are based on experience from the FRG approach, working with farmers, not with pastoralists and agro-pastoralists. It is in essence the same approach. However, it is likely that there will be further refinement of the guidelines based on new evidence and experience when the approach is used on the ground with pastoralists and other stakeholders.

The Ministry of Agriculture (MoA) is promoting group action in agricultural extension services. The intention is for farmers to organize at community level in a manner where innovative farmers can influence others to ensure better results. This is often called a Farmers' Development Group (FDG), which often consists of five to seven farmers living in near each other and led by one of the better-performing or more innovative farmers. In the promotion of FDGs, there is much interest in adapting the key principles of the FRG approach and development guidelines. Similarly, the second phase of the MoA's Agricultural Growth Program, which is supported by a consortium of donors, has adapted the promotion of farmer research extension groups (FREGs) in the target programme areas, mainly to enhance farmers' innovation. In all of these contexts, the developed guidelines are being used, with further refinement possible, based on the evidence gathered and lessons learned.

Training in the FRG approach

Design of the training programmes

Using the FRG research guidelines developed in FRG I, one of the main activities of FRG II was to bring the approach to nationwide attention by supporting training for the researchers at federal and regional agricultural research centres and universities with a faculty of agriculture. The training was organized for three groups of participants (Groups A, B and C), and each participant was expected to take all three stages of the training (see Figure 3.1). FRG II had trained 611 researchers, including 38 female researchers, by June 2014.

The overall objective of the training was to equip the researchers with an understanding of the FRG approach so that their participatory research activities and outputs would be improved. The expected output of the training was that researchers should be able to conduct FRG research activities or to apply the FRG approach to their existing and new activities. To reach over 500 researchers, the training was provided at six hub organizations in different parts of the country. The training featured components dealing with technical information, development skills, and gender sensitization workshop skills.

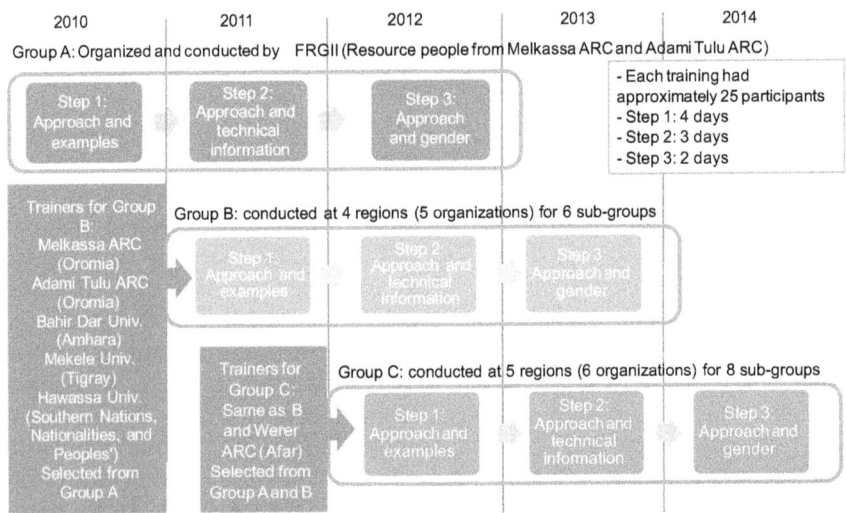

Figure 3.1 The framework of the FRG trainings

Facilitators and resource people

For the Group A participants, FRG II organized the training to make use of those who had experience in the FRG I research activities as facilitators. Among the Group A participants, FRG II selected potential resource people for the trainings for Groups B and C in the following years considering: -> ; 1) the capacity of the researchers: -> ; 2) the capacity of the organization to host and conduct the training; and 3) the geographical distribution and access within the regions to the hub of the region. The experienced researchers who worked with FRG I gave trainings to Group A. Then, the project monitored Group A participants for a year as they undertook the training and accumulated practical FRG experiences. Through this process, the selected researchers from Group A became qualified as trainers for the Group B and Group C training programmes. The training for the Group B participants was organized and conducted by the regional hub organizations, the Melkassa Agricultural Research Centre (MARC), Adami Tulu Agricultural Research Centre (ATARC), Mekelle University, Hawassa University and Bahir Dar University. For the training for Group C, Werer Agricultural Research Center (WARC) was added to the regional hubs, particularly for taking care of the training for the agricultural research centres and universities in Afar and Somali Region.

ATARC in Oromia Region and Bahir Dar University in Amhara Region hosted the training for two groups within Group B and C each, while the others hosted only one group in Group B and C. This was because Oromia and Amhara regions had more regional research centres and universities.

The training process

The facilitators and venues for the training for Groups B and C (Figure 3.1) were selected from Group A training participants after they had taken the basic training. They were a year ahead of Group B and two years ahead of Group C participants in ... in the three steps of training. They also accumulated practical FRG research skills by actually practising FRG research activities during the period between when they took the basic training and when they acted as facilitators training others. FRG II and candidate hub organizations, as well as the regional agricultural research institutes (RARIs), held discussions before starting any collaboration, shared the aims and objectives of the project, then jointly signed a memorandum of understanding. The selected trainers from all hub organizations met to discuss what they were going to do and how to conduct the training of trainers (ToT).

In the ToT sessions, the facilitators discussed the training programme contents and developed a presentation to be used consistently across all the training hubs. After sharing the common format of the programme and presentations, any fine-tuning was done by each training partner at the respective hubs.

Each hub sent an invitation letter to the potential participants from research centres and universities and handled all the necessary logistics for conducting the training. The training was observed and monitored by FRG II advisers from Japan International Cooperation Agency (JICA) and core team members consisting of Ethiopian researchers who had more experience in the FRG approach.

By exchanging information with the observers, the hub organizations absorbed some of the knowledge and experience necessary to successfully host the FRG training.

During subsequent review meetings of the training, the facilitators once again discussed what they had observed on the programmes, including individual components, time allocations and overall arrangements. They also shared the amendments they had made on the presentations. In addition, they shared information of what kind of comments and questions had been raised by the participants and how these had been dealt with during the training.

Content of the training

Step 1 of the training:

The first training sessions introduced the basic concepts of the FRG approach by going through the main features of the methodology shown in the 'Guideline to Participatory Agricultural Research through Farmer Research Groups (FRG) for Agricultural Researchers' (Bedru et al., 2009). Besides this, other main features covered were:
- a presentation about actual FRG research activities, followed by discussions;
- a field visit to FRG farmers;
- an FRG research proposal development exercise, plus presentation and discussion on already developed proposals, using information and experience accumulated during the training.

Step 2 of the training comprised:

- two presentations by two of the participants on their experiences in FRG or other relevant participatory research, followed by discussions;
- group discussion and presentations on issues common to all participatory research activities, such as how to conduct a field day, how to conduct an exchange visit, and tips to work with farmers;
- group work and presentations on how to facilitate wives' participation in FRG research; how data should be collected and how to maintain its scientific quality; how to involve the development agents in FRG activities;
- a field visit to FRG farmers;
- an introduction to a simple technical information presentation (Process Description Method) and follow-up exercise.

Finally, Step 3 of the training comprised:

- two presentations (one focusing on how to involve female household members in research activities and the other focusing on the differences between demonstration (as on-farm extension activities), scaling up (mainly focusing on the distribution of farm inputs such as fertilizers and seeds of improved varieties), and participatory research) by two of the participants based on their actual experience in FRG or similar participatory research, followed by discussions;
- an introduction to understanding and running a gender sensitization workshop;
- a preparation session for the content of a gender sensitization workshop, including subjects such as access to and control over resources, the daily activity calendar and division of labour;
- a gender sensitization workshop (access to and control over resources, daily activity calendar, division of labour and action plan development) with both male and female farmers.

Additional support for capacity development

The facilitators and trainers from the FRG approach training hubs, Ethiopian Institute of Agricultural Research (EIAR) and RARIs, were invited to Japan for experience sharing on participatory research each year during the FRG II project period. They discussed practical cases of participatory research with Japanese research and extension organizations, such as the Japan International Research Centre of Agricultural Sciences, Tokyo University of Agriculture, Nagoya University and Kyoto University, and visited some participatory research and extension sites.

In addition, FRG II conducted seminars on the participatory research approach in Ethiopia throughout the project period, and for such events the trainers were invited to share their experiences as well as to learn from the other researchers' experiences through real applications of the approach.

Through all these experiences abroad and in Ethiopia, the trainers added to their knowledge in participatory research approaches, helping to establish confidence throughout the hub organizations.

Challenges for the training hub organizations

Frequent staff turnover in the research centres and universities was a major challenge. Within six training hub organizations, the staff turnover or transfer of resource people was experienced at four organizations between 2011 and 2014. Moreover, previously trained researchers were often replaced by new staff who were not experienced in the FRG approach.

Various measures were taken by FRG II to cope with the frequent turnover of FRG II staff and facilitate consistent and effective training.

First, the project tried to organize a team of researchers as facilitators. For the training of trainers, at least three facilitators were invited to attend training courses if other staff were missing. With regard to the attendance of new researchers in the advanced training, they were trained by colleagues who had participated in previous training and were provided with training materials.

Team organization was flexible and left to the leaders or the facilitators within each hub, since interpersonal relations within the teams are very important. Also, to make the arrangements easier for team leaders, communication among the project and the focal people was normally shared with the responsible personnel of either of the faculty or research institute.

FRG II kept good communication with such management staff on a daily basis to provide timely support to the hub organizations. This also helped the institutionalization of the FRG training within the organizations.

Achievements of the FRG II training

FRG II managed to reach significant numbers of researchers across eight regions. In no small measure this was because most training hub organizations took ownership of the training and conducted the training programmes on their own initiative.

In addition, the FRG II project created a critical mass of facilitators who, while not involved in FRG I, understood the approach and could apply it very well, to the extent of training other researchers in different organizations.

The MoA has also started implementing the FRG/FREG approach across regions through its institutional network, focusing on different commodities, such as wheat, rice, cassava and dairy, which underlines the successful institutionalization of the approach.

FRG II support for research activities

In parallel with the training in the approach, FRG II has supported 43 research activities both financially and technically since 2010. These were located in eight regions. The topics of the research covered: rice-related technology (11 research activities), quality seed (seven activities), seed treatments (four short-term and three long-term research activities), farmer-saved seeds (two activities), irrigated vegetables (two activities), FRG at Farmer Training Centres-FTCs (four activities), forage-related technology in pastoralist areas (four activities), and other appropriate technologies (six research activities). FRG II not only supported the research activities financially, but also by improving research plans and reports and giving technical advice on-site and at review meetings, conducting training seminars, workshops and exchange visits in the participatory research approach as well as relevant technical topics. The support process followed the steps shown in Figure 3.2.

FRG IMPLEMENTATION: GUIDELINES, TRAINING AND RESEARCH 33

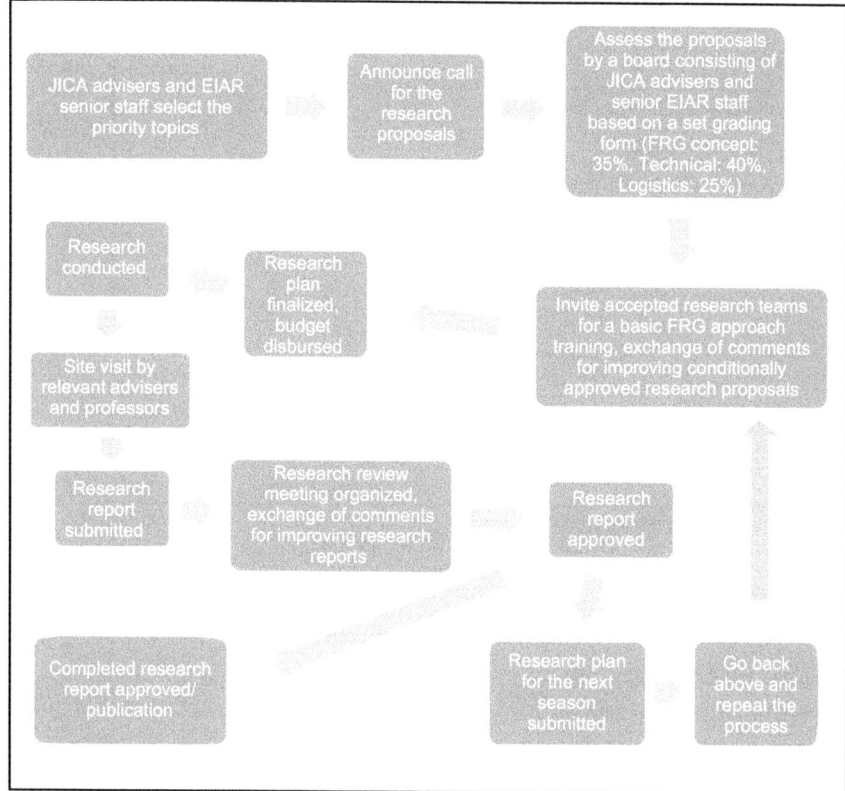

Figure 3.2 The support process for FRG research activities

FRG II has provided various kinds of support with the aim of improving researchers' scientific and participatory method skills, and also supported the publication of research reports, and enabled principal investigators to present their research in relevant seminars and workshops.

Research activity implementation

Identification of priority research thematic areas. The FRG II project identified six important thematic areas for research. These are: 1) rice cultivation technology development, 2) quality seed technology, 3) irrigated vegetable production, 4) forage technology development, 5) FRG research at farmers' training centres (FTCs), and 6) other agronomy-related research activities. FTCs are staffed by development agents and established at the village level nationwide by the Ethiopian Government to train farmers, demonstrate new technologies, and deliver necessary information. The research projects that have been selected and supported by the project are summarized in Table 3.1.

Table 3.1 Thematic areas and number of research projects supported, 2010–14

Thematic area	No. of projects supported	No. of research organizations involved
Rice cultivation technology development	11	8
Quality seed technology	16	11
Irrigated vegetable production	2	2
Forage technology development	4	4
FRG research at farmers' training centres	4	4
Other agronomy - related research	6	6

Table 3.2 Number of research proposals submitted and selected for support 2010–12

Thematic area	No. of proposals submitted (year)	Number of proposals selected (year)
Rice cultivation technology development	11 (2010) 8 (2011)	6 (2010) 4 (2011)
Quality seed technology	1 (2010) 15 (2010–12) 33 (2011)	1 (2010) 4 (2010–12) 8 (2011)
Irrigated vegetable production	3 (2011)	2 (2011)
Forage technology development	22 (2011)	4 (2011)
FRG researchat farmers' training centres	139 (2011)	4 (2011)
Other agronomy-related research	222 (2012)	6 (2012)

Call for proposals. To select feasible research projects, a call for proposals was issued via various channels. This included website, fax and email announcements to all agricultural research institutes, their centres, and universities. This approach enabled the largest number of researchers to apply from different research institutes. Though responses varied according to the theme and particular year, a sufficient number of proposals were submitted for meaningful comparison and judgement to be made. The largest number of proposals submitted was in 2012 for agronomy-related research – 222 for six research grants (see Table 3.2).

Evaluation and selection of proposals. The submitted proposals went through a two-step review process. The first step was a pre-screening by FRG II staff based on a simple consideration of whether the proposal met the required components as outlined in the proposal submission template. In the second step, a review panel of senior researchers in their respective fields of specialization evaluated each proposal by considering: 1) relevance of the research and its potential impact on small-scale farmers, 2) the appropriateness of the experimental design in terms of treatments, methods of data collection and analysis, and consideration of scientific value addition, 3) the appropriateness of the approach proposed for implementation and evaluation of the research activity, and ultimate delivery of its outputs to the target farmers/stakeholders, 4) the technical capacity of the proposed research team, and 5) whether the budget and timeline proposed were appropriate and achievable.

Review meeting: upgrading selected proposals. Before funding the selected proposals, the project organized a review meeting, inviting three research team members from the selected proposals and senior researchers to review specifically for possible improvement. The experience from these reviews showed that it has been a key instrument in improving the design and implementation of particular research activities, particularly regarding methods of data collection and analysis.

Technical 'backstopping', monitoring and evaluation. There are regular monitoring and evaluation procedures throughout the implementation period, including: 1) research site visits by project staff, 2) regular reporting on the status of the respective research activities, and 3) engaging selected members of the research teams in other activities such as training on developing extension materials, the theory and practice of participatory research, and approaches and principles involved in the FRG approach. This programme creates opportunities to follow the progress of the research activities, develop skills and perform a 'backstopping' function if problems develop.

Final research report review meetings. The research activities often have a two to three-year implementation period. Once the activities are completed, the respective research teams are expected to produce a scientific research report and present the programmes of completed activities for each year of the project at a review. In these review meetings, in addition to the project staff who have been running the research activities, senior researchers with relevant areas of specialization are invited. The key purpose of the review is to evaluate: 1) the overall scientific consistency and coherence of the report, 2) the appropriate application of the FRG approach, including a detailed description of farmer participation, 3) the appropriateness of the analytical method(s) applied, 4) the appropriateness of the inferences made from the research analysis, and 5) the appropriateness of the research and development recommendations made. The researchers are then required to revise the research report, based on the suggestions and comments provided during this review process. The research report is considered the principal FRG project output only after all of the suggestions for improvement have been addressed.

Revised research report and raw data submission. This is the final step, where research teams are requested to submit the revised research report together with the raw research data. Before publishing the research report, the submitted reports are reviewed further by editors, often senior researchers with relevant expertise in research and scientific writing. In some cases, checking the raw data for consistency is required.

Sustainable support for research activities

FRG II has supported FRG-based research activities both financially and technically, to provide opportunities for agricultural researchers to try out the approach and to collect the necessary feedback to adjust the FRG approach in different agro-ecological, socio-economic and cultural settings. Participatory research activities had been conducted within the regular research framework in Ethiopia even before the initiation of FRG I. However, most of the research was carried out more as extension-oriented activities, rather than producing new technical information as a research output.

The support of research activities by FRG II was designed with consideration for the existing government framework for research activities. The budget for each research activity was set at ETB 25,000 Ethiopian birr (USD 1,942.50) a year per research project in 2010, and ETB 50,000(USD 2,590) a year per project in 2014. The amount was set following the amount of budget normally allocated for a participatory research activity funded by EIAR and RARIs, taking into consideration the sustainability. FRG II designed the support processes and sustainable budget for each research project, so that the activities supported by FRG II could be easily supported by either federal or regional agricultural research institutes and could continue even after FRG II funding finished.

The subjects given out in the call for research proposals were decided bearing in mind the existing research priorities of Ethiopian agricultural research institutions and JICA, so that additional technical support and collaboration with other JICA projects could be facilitated. Also, it was expected that collaborator activity would assist with the task of bridging the research–development linkage. In each case, when the FRG II project selected and announced the research topics, JICA advisers and senior researchers of EIAR discussed how to make sure the topic would meet the priorities set by the Ethiopian Government, and be able to make use of the technical support offered by Japanese technical cooperation projects.

FRG II collaborated with and was supported by different JICA technical cooperation projects on different research topics. In relation to the 'Quality seed production technology' which also included 'seed treatment' and 'evaluation of farmer, saved seed for quality in teff/ wheat', Quality Seed Promotion Project (QSPP), which was a technical cooperation project between the MoA and JICA, support was given in terms of the topic selection and providing related technical information. The proposal was limited to target sites in either the

Sogido-Saraweyba Irrigation scheme or Hirna Irrigation scheme where Project for Capacity Building in Irrigation Development (CBID) was already working on capacity building for regional experts, the rehabilitation of the irrigation structure, and organization of the water users' associations. CBID hugely supported the FRG research in terms of monitoring research activities for vegetables under irrigation. For the research activities of forage improvement in pastoral and agro-pastoral areas, the Rural Resilience Enhancement Project (RREP) provided essential background information on the pastoralist community in the Borena area. In addition, FRG II tried to work with the Deutsche Gesellschaft für Internationale Zusammenarbeit (GiZ) Sustainable Land Management Project on FRG@FTC. However, it proved difficult to encourage the researchers around the FTCs supported by GiZ. The call was instead sent out to any researchers under NARS instead.

The whole process of FRG-based research activities supported by the project followed the routine process of regular research activities conducted by EIAR, RARIs and universities. At the selection process stage, all of the proposals submitted were evaluated by JICA advisers to FRG II and relevant senior researchers of EIAR, following a scoring sheet checking the participatory, technical and logistical aspects. The geological distribution of projects was also considered, because one of the main aims of FRG II was to promote the FRG approach nationwide. The result of the selection process was reported to the principal investigators, who led the teams of multidisciplinary researchers to work on FRG research activities, and their superiors, and two other research team members were invited to the FRG training in case of staff changes. After the training, each research team was requested to comment on and improve the research proposal based on the principles of the FRG approach.

Once the research proposals were finalized, the implementation process followed the regular routine at EIAR. This begins with the research budget disbursement to the official account of the relevant research organizations. For monitoring, JICA FRG II advisers and other relevant facilitators visited the research activities (27 visits in the 2013 season for 16 research activities). After each cropping season, the research teams were requested to submit the research report and present their output at review meetings, in line with the usual processes of EIAR, RARIs and universities. In the review meetings, the research teams collected comments and were given assignments to work on the improvement of reports until they were approved. Again, based on the research outputs, next season's research plans were offered for comment and the research teams requested to work on improving them. This whole process was repeated every season.

Challenges in supporting the research activities

As of June 2014, 17 research activities had been completed, 17 had been terminated, and nine we on going, out of 43 activities. The following were some of the main reasons for terminating support by FRG II:

1. Staff turnover.
2. When the principal investigator transferred to another role, information on the research activities was not properly handed over to the new principal investigator.
3. The research report was missing necessary data and could not be brought to an acceptable level.
4. Field research management, the experiment was devastated and left in no fit state to collect data (the crop failed for two consecutive years in all the trial farmer's fields).
5. The research topic required further testing on-station.
6. The research report could not be improved to an acceptable level before the second season sowing.

Among these reasons, 1 and 2 were the main reasons that FRG II was not able to continue support for the research activity. However, FRG II did not terminate the support without much consideration. The above-mentioned 17 research activities were only terminated after a series of communications trying to address the respective problems in each case. Thirteen out of the 17 terminated activities did not have sufficient data to improve upon, despite repeated efforts. There seemed to be two different reasons for this.

The first was that the research report was not accurate enough, because of decisions made by the researchers during the experiment. This usually came to light during monitoring visits. When the sites were visited, the support team found something different happening from what had been agreed in the research proposals. There was an agreed process to follow in case any such changes needed to be made. Usually, however, the research team had changed the initial approved research plan without consultation, or fully considered scientific justification. By the time the support team visited and observed the trial, it was too late to return it to a state where it could be a scientifically valid research project.

The other main reason for abandoning a trial was that even though the experiment itself was conducted without a problem and with the agreed design, the researchers failed to collect sufficient and appropriate data to prove or reject their hypothesis. Everything seemed to be going fine on the trial site during monitoring visits, and although the researchers seemed to be aware of what was necessary when their reports were received, they were missing essential data and the evidence required to prove their statements. There seems to be two main causes for this: firstly, the researchers' limited skills in data collection, and secondly, the design of the research activity was not refined sufficiently. The critical importance of primary/raw data is not always well understood by researchers. Thus, by the time a problem is discovered in the analysis, it is too late to go back to the raw data to correct it. Either the data is inadequate, or in some cases the researchers had discarded the raw data immediately after they entered it into the relevant software.

In addition, about half of the researchers did not respond to the comments given during review meetings. Despite statements such as 'We appreciate the constructive comments when they were made at review meetings. All the comments were accepted and we will make improvements reflecting those', in our experience, many of the researchers either did not understand the comments or ignored them.

Possible causes of challenges

We can summarize the challenges experienced into three types: 1) limited scientific accuracy in conducting an on-farm trial; 2) not presenting the necessary evidence in research reports; and 3) not responding to the review comments in the research reports.

We have observed many cases where no report, communication or consultation with the appropriate people at the appropriate time took place. When something unexpected happened at the trial sites (common in participatory on-farm trials), the researchers were usually requested to make a decision on the spot. In many cases, the researchers on-site made decisions without thinking of the consequences, or thinking that it wouldn't make a significant difference to the research output. The essence of the FRG approach is the complementary skills of the multidisciplinary research team. This is intended to assure the capacity of the researchers to respond to such 'unexpected' circumstances, enabling them to discuss the best approach among themselves, and then ask for help if needed. In cases where, in spite of repeated requests to make contact with FRG II advisers or senior researchers when there were unexpected changes, and possible amendments needed to the research design, if contact still didn't happen, the research project was likely to fail at that point.

The second example, where there is missing evidence in the research report, comes possibly from a misunderstanding about participatory research projects with promotion/extension activities. In addition, we often observed that many researchers presented their opinions without supporting data/evidence. It is essential for researchers to collect evidence and discuss something based on that evidence, whatever type of research activity they are engaged in. Some researchers, on the other hand, included unnecessary data and analysis to the research report, without considering whether it was essential to support their statements. A clear grasp of the principles of scientific report writing is required and clearly has been missing in some cases. Finally, the basic purpose and objective of the analysis must not be overlooked.

In cases where supporting data was missing, we advised researchers to collect the required data during the following season. Some followed this advice and the reports were improved, while others repeated the same mistake. The common root cause for these difficulties again was inexperience and lack of knowledge about the practical methods of data collection by junior researchers.

In the case of the third challenge – not responding to the given comments on the research report – this was caused by there not being a clear process embedded in the routine of EIARs and RARIs to follow up the discussions that had taken place in the review meetings. Since there were no checks carried out by anyone to see if the review meeting comments were being worked on, the researchers in some cases did not take responsibility for making the necessary improvements. We also discovered that researchers' inexperience working with Microsoft Word also meant that (despite frequent reminders from FRG II project staff) there were problems with working on the latest version of a report, or even using 'track changes', so it was unclear which parts of the report had been amended. This problem could also have been addressed if there had been a system of follow-up checks, as mentioned above, and training for junior researchers in using the required software, with support from their senior staff.

As for addressing the challenge of improving the quality of the report, we tried to check the draft report very carefully, sentence by sentence, as the writer responded to the given questions and comments. We did not approve next year's activity unless the research reports were improved to the appropriate standard. Finally, we tried to make comments as clear and constructive as possible to guide the writers to improve the report's content, rather than simply making negative critical observations.

Conclusion

The guidelines as developed clarified the methodology and set the agenda for understanding the FRG approach for agricultural researchers. Researchers are now using the guidelines, applying the essence of the approach in the reality of farmers' fields in different agro-ecological and socio-economic contexts. It is a sensible use of the guidelines because in participatory research there is a huge diversity of circumstances in which to apply the approach, depending on the target commodity, farmers, their communities and geographical locations.

The effectiveness of the approach is now recognized and is applied in many and various activities related to agriculture, and there is a policy of continual improvement of the guidelines. Continuous feedback from the field is seen as essential.

Regarding training on the FRG approach, the majority of the training in Ethiopia has been focused on explaining what the FRG approach is and creating awareness. However, there is recognition that to actually improve the quality of research activities, it is increasingly necessary to shift the priority towards 'how to do things better'. The initiative for this shift can already be observed in the government system in different sectors. Time and effort was invested to start the knowledge and experience transfer about FRG, and to gain momentum. The programme is now rolling, and the coming challenge is to keep the momentum.

There are two major remaining challenges in relation to institutionalization, which are both related to the creation of functional information-sharing/reporting systems and externalization of the work among researchers. On the first point some organizations took more than a month to submit the simple report and related information after conducting the training, despite repeated requests. This seems to be representative of the current status of information sharing and reporting within the relevant organizations. Also, the necessary arrangements to carry out the training programme were not well-managed, in that the resource people still had to prepare for the next programme while working on on going activities. This is also thought to be related to a lack of information-sharing/reporting systems. The second issue is that although the training was conducted following the memorandum of understanding between FRG II and each organization that the training would be run as one of their own activities, many of the staff requested additional arrangements for work associated with FRG training and some refused to work without additional incentives, feeling that it was extra work out of their normal responsibilities.

There are two critical remaining issues in the agricultural research system in Ethiopia which do not only apply to FRG research, but also to regular research activities including conventional on-station research activities. The first is reluctance to engage in reporting, communication and consultation. The researchers need to be trained to do the required research and associated activities, beginning as students. In many cases, reporting, communication and consultation in the early stages of a problem can resolve it easily. However, in many cases, the researchers seem to feel embarrassed to ask questions or ask for technical advice. This is especially important for junior researchers because, by doing so, they can learn the necessary professional and practical knowledge from senior researchers.

In addition, the capacity of junior researchers is not improving at as fast a rate as anticipated, mainly because 1) the knowledge of how to ensure scientific accuracy and validity in participatory research is limited among researchers generally, and 2) the follow-up after review meetings about the quality of research outputs is often not as expected. This means that junior researchers miss precious opportunities to improve. Moreover, since many research reports are sent somewhere else once finalized, the data are not accumulated at research institutes as a repository of scientific evidence available within those organizations.

Overcoming these challenges is not straightforward. However, by addressing them, the FRG approach can be effectively, consistently and appropriately applied, and will achieve results that will lead directly to improvements in farmers' livelihoods.

About the author

Taku Seo works with the Smallholder Horticulture Empowerment and Promotion Project for Local and Up-scaling (SHEP-PLUS), a technical cooperation project between the Ministry of Agriculture, Livestock and Fisheries of Kenya and the Japan International Cooperation Agency. He worked for FRG II between 2010 and 2015 as an expert in charge of training and appropriate technology development.

References

Bedru B., Berhanu S., Endeshaw H., I. Matsumoto, M. Niioka, K. Shiratori, Teha M. And Wole, K. (2009) *Guideline to Participatory Agricultural Research through Farmer Research Group (FRG) for Agricultural Researchers*. Project on Strengthening Technology Development, Verification, Transfer and Adoption through Farmers Research Groups (FRG I), Addis Ababa: EIAR/OARI/JICA.

PART II
Experiences of Farmer Research Groups

CHAPTER 4
Engaging farmers in technology evaluation and promotion: Farmer Research Groups on common beans in the Central Rift Valley, Ethiopia

*Endeshaw Habte, Kidane Tumsa,
Berhanu Amsalu Fenta and Abiy Tilahun*

Abstract

This research is part of an attempt to contribute to the understanding of the needs and interests of small farmers by way of working closely with them. It was carried out with Farmer Research Groups in three districts to identify the most suitable varieties of common beans for specific locations. The variety selection/evaluation process was carried out using farmers' own criteria: utilization, marketing and field performance. The results indicated that the selection criteria of farmers in the three locations were very similar except in very few cases where they varied in the emphasis given to a particular criterion. Accordingly, the varieties that suit farmers' respective locations were identified. On-farm participatory technology evaluation and transfer with well-organized farmer groups working together with research and extension teams has proved a valuable interactive approach for rapid technology evaluation and dissemination.

Keywords: participatory research, technology evaluation, farmer preference, Farmer Research Groups, common beans, Ethiopia

Introduction

The benefits of agricultural research and development efforts in developing countries 'sometimes reach only a limited number of people, have a minimal impact on poverty, and are difficult to sustain over the long term' (FAO, 2007: 1). Differences among target groups (communities) often stand as barriers to the use and spread of useful agricultural technologies. Farmers in developing countries tend to be smallholders, diverse, and operate in complex socio-economic as well as biophysical environments (Ellis, 1993). Meeting their needs, therefore, requires a proper understanding of their social and economic contexts as well as relevant institutional settings. Ignoring these differences under the premise of homogeneity leads to lower levels of adoption of technologies 'developed for them' (Feder et al., 1985). This in turn underlines the fact that most technology adoption studies are inspired by poor technology adoption by farmers, despite many development interventions aimed at easing constraints such as those associated with finance and access to information. Interventions need to be tailored carefully to the circumstances of the target farming community if investments in research and extension (from limited resources) are to be effective.

Finding appropriate technological solutions in the context of small farmers demands 'better identification of practices that have demonstrated economic, social and environmental benefits at the community level, together with policies and programs that support the spread of these practices' (FAO, 2007). Such practices and operational frameworks require building and sustaining a purposeful partnership at grassroots level. Participatory research and development approaches are useful instruments in customizing technologies to fit local realities (Willem and Bertus, 2004).

Participatory technology development has been advocated as a research and development process applicable to resource-poor, marginal and complex farming systems such as those found in semi-arid areas of sub-Saharan Africa. Empowerment of the participants, increased confidence of farmers and local people in their own knowledge, improved capacity of clients to innovate and experiment, and an enhanced ability to cope with change are often claimed to be achievable using participatory methods rather than through traditional technology transfer (Mellis et al., 1999). The past two decades have seen an increasing recognition of the importance of participation by beneficiaries and a wide range of stakeholders in decision-making. This has led to the development of various participatory approaches, tools and methods that facilitate innovation development and application in diverse contexts. Experience has shown that participation improves the quality, effectiveness and sustainability of development actions (Ponniah et al., 2008).

The Ethiopian agricultural research system has promoted participatory research to address farmers' biological and socio-economic constraints by developing and scaling up technologies with the active involvement of farmers. Encouraging results have been observed in the process, particularly by improving interaction among stakeholders. This has in turn shown a need to further improve and institutionalize the participatory approach in the research system for rapid and tangible research impacts. Owing to this, a project on strengthening technology development, verification, transfer and adoption through Farmer Research Groups (FRG I) was jointly launched in 2004 by the Ethiopian Institute of Agricultural Research and the Melkassa Agricultural Research Centre (EIAR/MARC), Oromia Agricultural Research Institute and the Adami Tulu Agricultural Research Centre (OARI/ATARC), with assistance from the Japan International Cooperation Agency (JICA). The project was conducted in the East Shewa Zone of Oromia Region. The main purpose of the project was to establish FRGs as one of the research and extension approaches in the project area and to extend the experience to many other areas in the country.

Among the commodities the project has promoted through participatory research is the common bean (*Phaseoulus vulgaris L.*). It is one of the lowland pulse crops produced in various parts of Ethiopia, including central, southern, eastern, western, north western and northern regions. It is predominantly grown for cash in the Central Rift Valley (CRV) and also for food as a supplementary protein source for poor farmers. The most common and popular traditional dish prepared from common beans in the CRV is *nifro* (Yetneberk, 1995). Early maturity and double cropping often make beans the first food to become available after the annual 'food supply gap' and sometimes the only crop to survive in a short growing season (Nigatu et al., 1995).

The common bean has been accorded a high priority because of its significance in domestic consumption and as an export crop. Farmers sell dry beans, fresh beans, immature pods and dry bean flour in food markets throughout Ethiopia. The lion's share of export production comes from small farmers, especially in the CRV, followed by southern, eastern and western regions of the country (Abate, 1995). In 2005 the country earned USD 60 million from exporting these beans (Ethiopian Custom Authority, 2006).

The Central Rift Valley presents a particular opportunity and challenge to common bean production and marketing. It contributes 60 per cent of production and accounts for 40 per cent of the total area covered with beans in the country (Nigatu et al., 1995). Until recently (mid-2000s), common beans remained an 'orphan' crop with farmers paying it little or no attention. Weeding, line planting, proper land preparation and other management practices were hardly considered, and the production purpose did not go beyond local consumption and a source of cash (as it matures early) to meet labour and related expenses for the harvesting and threshing of other major crops, such as maize and teff. Thus, productivity in the CRV falls within the range of 300–800 kilograms per hectare, well below the regional (1000 kg/ha) and national (900 kg/ha) average (Asefa et al., 2006).

The national lowland pulse research programme in Ethiopia developed and released a diverse set of common bean varieties with wider and specific adaptations. Some were meant for market and others for local consumption (see Table 4.1). The varieties have various qualities such as better yield, a shorter maturing cycle, improved cooking qualities and market niche value, compared to the ones farmers were already growing. This was an opportunity for the CRV farmers, who often have to wait too long either to feed their families or to earn enough money to meet their running costs for harvesting and threshing their other crops. Yet, most of the bean growers in the CRV were using only their poor-yielding cultivars as well as a limited number (usually one or two) of older improved varieties, mainly because they had little opportunity of coming into contact with other available options to widen their choice of what to grow.

Table 4.1 Important features of the bean varieties used for the participatory evaluation with FRG farmers, 2005

Variety	Released year	Days to flowering	Days to maturity	Seed colour	Seed shape	Seed size	Yield (kg/ha)	Potentially suitable areas	Use
Awash-1	1989	38	75	White	Round	Small	2,400	CRV	Canning (export type)
Roba-1	1989	42	75	Cream	Elongated	Small	2,100	All locations	Food, especially soup, *shiro*, *kike*, samosa and *nefro* (boiled)
Mexican-142	1974	56	86	White	Round	Small	2,100	All locations	Canning (export type)
Atendaba	1996	41	74	Cream carioca	Round	Medium	2,400	CRV	Food especially for *nefro* (boiled)
Awash Melka	1998	42	75	White	Round	Small	2,500	CRV	Export type
Zebra	1998	38	76	Cream carioca	Elongated	Medium	2,700	CRV	Food use as *nefro* (boiled)
Tabor	1996	42	76	Cream	Elongated	Small	2,000 – 2,500	Southern region (around Awassa area)	Food use with *kocho* and *gomen*
Ayenew	1996	41	77	Cream	Round	Large	2,500 – 3,000	Eastern region (Hararge area)	Food use as *nefro* (boiled) with sorghum
Gofta	1996	36	78	Cream	Round	Medium	2,500 – 3,000	Eastern region (Hararge area)	Food use as *nefro* (boiled) with sorghum
Brown speckled	1974	38	78	Cream	Elongated	Large	1,100	All locations	Food use as *nefro*
Beshbesh	1997	42	76	Cream	Round	Small	2,500 – 3,000	Specific for southern region (Wolaita area), for bean fly resistance	Food use as *nefro* (boiled) with *kocho* and *gomen*

(continued)

Table 4.1 Important features of the bean varieties used for the participatory evaluation with FRG farmers, 2005 (continued)

Variety	Released year	Days to flowering	Days to maturity	Seed colour	Seed shape	Seed size	Yield (kg/ha)	Potentially suitable areas	Use
Goberasha	1998	38	78	Red	Elongated	Large	2,500 – 3,000	Western region around Jimma and Wollega area	Food use as *wot* (sauce) and *nefro* (boiled)
Red Wolaita	1974	42	77	Red	Elongated	Small	1,400	All locations	Food use with *kocho*, *gomen* and *nefro* (boiled)
Melke	1997	38	78	Red	Elongated	Large	2000 – 2,500	Specific for southern region (Wolaita area), for bean fly resistance	Food use as *nefro* (boiled) with *kocho* and *gomen*
Dimtu	2003	61	86	Red	Elongated	Medium	2,200 –2,400	CRV and Southern Ethiopia	Food used as *nefro* (boiled) with cereals and *kocho* as *wot* and *wasa*
Nasir	2003	61	88	Red	Round	Medium	2,300–2,700	All locations	Food used as *nefro* (boiled) with cereals, *wasa* with *kocho* and *wot*

Note: CRV = Central Rift Valley

This chapter presents the experience of applying the FRG approach in evaluating and promoting different varieties of common beans in selected districts of the Central Rift Valley of Ethiopia.

Research methodology

Common bean varieties considered for evaluation and promotion

In the districts where the participatory on-farm evaluation activity was carried out, the common bean is among the most important crops, second only to maize, being grown both as food and as a cash crop. Though maize growing dominates, common beans are recognized as a security crop in times of climatic uncertainty, that is, when farmers face problems planting maize due to the late onset of rain. Beans then become the only alternative. Despite being a valuable crop, use of improved varieties and management practices were not common. Besides, district agricultural extension workers had limited knowledge about the technological options available with research.

Establishment of Farmer Research Groups

Owing to the gap in knowledge and practice identified in the Central Rift Valley researchers from MARC took a step towards applying/using the Farmer Resarch Group approach in technology evaluation and diffusion, with the aim of identifying the most appropriate technical options (bean varieties) for specific locations, improving their performance and widening access to the technologies that met the standards of farmers' selection criteria. This joint research and extension activity was shaped by the following question: how can the FRG approach be made an interactive platform for research, extension, farmers and other stakeholders in the technology validation and diffusion process?

The research activity was carried out for two years (2005–06). The first year focused mainly on evaluation and identifying the best varieties for the respective sites, while in the second year farmer-based seed multiplication of the best (preferred) varieties was carried out to improve availability and access to seeds of those varieties.

In carrying out the activity, the FRGs that were already working on other participatory research topics in the three sites were used. Initially, membership of the group was determined based on farmers' interest in being involved and their voluntary commitment to share part of the cost (e.g. land, labour) for the research activity. There were 14 member farmers (farm households) in Bora, 15 in Shala and 16 in Adami Tulu Jido Kombolcha (ATJK) districts respectively. The majority of group members were male farmers. The number of female farmers in the group ranged from two each in Bora and Shala to one in ATJK district. The district agricultural office experts and development agents as well as a multidisciplinary research team composed of a social scientist (agricultural extensionist) and natural scientists (breeder, pathologist, entomologist, agronomist, agricultural engineer and food scientist) jointly worked with the FRGs in the implementation process.

Participatory evaluation of technical options

Starting the discussion with farmers. The entry point for the common bean trial was a focused group discussion with FRG farmers and respective district extension offices in the selected districts. During these discussions, farmers and extension workers reported that bean cropping was constrained by poor-quality seed, little or no use of fertilizer, no weeding carried out because of less attention given to the crop, and the low market price. A lack of awareness of improved management practices for common bean production was also mentioned.

Possible remedies were suggested during the discussion, including the opportunity to learn about improved technical options, exchange visits to research centres and to other progressive bean farmers to learn about improved practices the provision of timely market information, strengthening the supply networks of the best varieties, and training for farmers and extension workers. It was agreed to begin by jointly testing the varieties available and identifying any that were better than those available to farmers, and at the same time working out ways that improved technologies could be accessed locally.

Before research activity began, there were frequent visits to respective FRG sites to introduce and explain the FRG approach. The discussions included the role of each member of the group (farmer, development agent, researcher, subject matter specialist) in the evaluation, improvement and dissemination of technologies. We also discussed the requirements of being an FRG member (willingness and commitment) and the significance of the FRG to other farmers in the area.

The FRG approach in shaping the experiment. Following the discussions, the research team identified 15 different varieties available in the national bean research programme (during 2005). They were, namely, Nasir, Brown speckled, Awash-1, Atendaba, Mexican-142, Awash Melka, Dimtu, Roba-1, Beshbesh, Zebra, Ayenew, Tabor, Melke, Goberasha, and Gofta. To keep farmers involved and informed, the team organized a meeting to introduce the common bean varieties, their characteristics and management requirements, to both the FRG farmers and extension workers in the respective trial sites. Displays about each variety were arranged and printed information leaflets on common bean varieties and cultivation practices were also provided. During the session the criteria used to select common bean varieties were devised by the farmers themselves. These included seed colour, size, taste, cookability (cooking time and their state after cooking), marketability (demand in the local and/or international market), tolerance to pests and diseases, and yield performance.

While the physical and market characteristics were judged based on visual observations, other characteristics such as cookability and taste were evaluated after the FRG member farmers (female heads and wives of the male heads) had prepared the most common recipes using a few kilos from each of the 15 varieties provided for the purpose and brought to the evaluation session. The evaluation of physical, market, cooking and taste attributes of the varieties involved all the FRG member farmers (45 household heads, both male and female) and their spouses (26 wives). During the evaluation session, the farmers were asked to rate each variety as: *very good* (score 1); *good* (score 2); or *poor* (score 3), against each criterion. In a few exceptional cases, the rating *very poor* (score 4) was used. The evaluation and ranking of the varieties using farmers' criteria, prior to moving to the field trial, was a unique experience where male and female farmers were able to debate, interact and reach a consensus view. The varieties were then ranked based on the total score given against each of the criteria. Varieties with the lowest score assumed the top rank and when there was a tie (particularly with the top three), farmers were asked to discuss further and to differentiate the order of preference (see Table 4.1). This activity was designed to clearly identify the varieties to be field tested. However, the lower-ranked varieties were also tried, with some lesser follow-up, to observe if there had been any change in preferences due to the field performance of the variety.

Accordingly, it was agreed with the group to conduct the trial using the top three ranked varieties, replicated on four FRG farmers' sites. The trial farmers were identified based on the consensus reached among FRG farmers and extension workers regarding the best sites to use. However, in one of the districts the lottery method was used because reaching a consensus proved difficult. Participatory selection of farmers to host the trial is intended to avoid potential misconceptions that the research activity is targeted at just a few member farmers and it maintains the willingness of all the FRG farmers to be actively involved in the activity.

It was also agreed to provide a few kilos of the remaining varieties (ranked from fourth to last) to the remaining group members to observe the field performance of the lower-ranked varieties (see Figure 4.1). This was done to identify the superior variety, if any, out of the remaining 12 varieties, and also to keep the other group members involved in the trial process and group activity. The availability of so many bean varieties, in this particular case, provided a good opportunity to sustain the participatory process through engaging each group member in the experiment.

In total, there were 12 farmers that hosted the on-farm evaluation trial for the three top varieties, across three districts. These trial plots were closely followed up and the evaluation data was recorded, based on observation as well as discussion with FRG farmers.

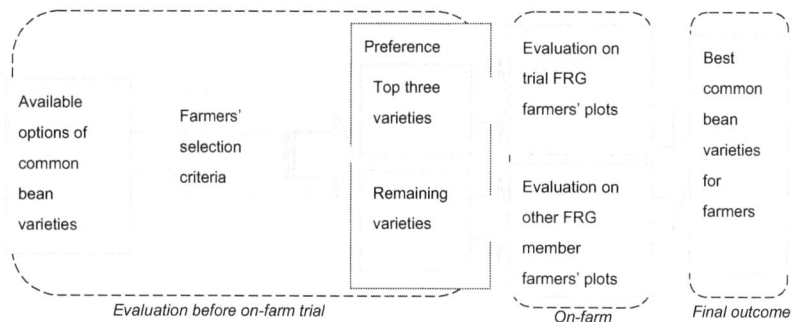

Figure 4.1 The participatory selection process of bean varieties

On-farm trials. The field trial was carried out on selected FRG farmers' fields in such a way that four varieties – the three top ranked improved varieties and one local variety (see Table 4.2) – were planted side by side on equal-sized plots (625 square metres for each variety), replicated by the number of participant farmers (four farmers at each site).

Regular discussion sessions took place for FRG members, DAs and district agricultural office experts about the technical and related matters of the trial. The FRG farmers were, in general, given the role of carrying out the field activity on their own land as per the agreement reached at planning stage. The trial farmers also recorded their observations on the results. Other member farmers who received the lower-ranked varieties also did the same.

Table 4.2 Preferred common bean varieties for field testing

District (kebele)	Common bean varieties tested				No. of farmers
	Variety	Seed colour	Seed shape	Seed size	
Adami Tulu Jido Kombolcha (Horofalole)	Ayenew	Cream	Round	Large	4
	Dimtu	Red	Elongated	Medium	
	Goberasha	Red	Elongated	Large	
	Local	White	Round	Small	
Shala (Awaragema and Bekeledeya)	Brown speckled	Cream	Elongated	Large	4
	Awash-1	White	Round	Small	
	Nasir	Red	Round	Medium	
	Local	Red	Elongated	Small	
Bora (Bertasami)	Tabor	Cream	Elongated	Small	4
	Roba-1	Cream	Elongated	Small	
	Awash Melka	White	Round	Small	
	Local	White	Round	Small	

Based on observation by the research team, tailor-made classroom practical field training was organized to meet specific needs/gaps observed in the implementation and management of the activity (for example, how to distinguish disease and insect pest symptoms, how to record observations and so on). During the training sessions, the FRG farmers and extension workers discussed any problems or challenges that arose in relation to the field trials or the group itself, as well as how to handle future tasks.

Field visits were conducted once a fortnight with a team of researchers, DAs and district agricultural office experts. The district experts participated in scheduled meetings that were conducted at different stages: before planting, after planting, mid-season, at time of crop maturity and after the harvest (evaluation meeting). In addition, exchange visits (FRG to FRG) were arranged to provide inter-group learning and communication. The FRG farmers and district agricultural office experts from Bora and ATJK were taken to Shala where the trial farmers and agricultural experts performed well with the facilitation of group activities and individual management of trial activities. The process of nurturing innovative capacity is as important in farmer participatory research as the actual field results.

Important field data, using farmers' selected parameters, particularly regarding yield and tolerance to disease and insect pests, were collected from each plot. The field performance in terms of disease and insect pest resistance was measured based on standard scoring methods (a scale of 1 to 9 where 1 is highly resistant and 9 is highly susceptible to major diseases; similarly, a scale of 1 to 5 was used for insect pest scoring where 1 is resistant and 5 is susceptible to major insect pests). Overall plot/variety assessment/scoring was carried out based on visual evaluation with the involvement of the FRG farmers.

Following the evaluations made before and during the trial stage, an overall performance assessment was made for all the bean varieties using three important parameters: yield, marketable quality, and resistance to disease and pests. The marketable quality issue was considered important because farmers suggested that after a year of trialling, they were well-qualified to assess the market potential of the various beans, including those which had not been popular in local markets before the trials were carried out. This time the farmers' evaluations were made using the on-farm trial experience and visits made to each other's trial plots. In some cases, the rank which was set at the pre-planting stage remained, while in other cases the preference list was changed to include some of the varieties not previously in the top three but which had performed well in the trials. The results are summarized in Tables 4.5 and 4.6.

Extension event and seed production

Field visits were organized to promote interaction between FRG and non-FRG farmers, to create wider interest and awareness of the bean varieties and to share experiences. These were attended by the respective district agricultural and rural development office heads, agricultural specialists, researchers, DAs and the farmers around each of the trial sites. By the FRG farmers sharing the experience of the trials the farmers in the wider district got to know about the qualities of the various bean varieties, which in turn initiated farmers around the trial sites to seek access to and produce seeds of the varieties endorsed through the research process. In response to the demand created, farmers started their own seed production activities in the second year of the programme.

Results and discussion

Agronomic evaluation

The average seed yield figures collected in the ATJK district indicated that the highest yield was obtained from the Dimtu variety, followed by Goberasha and Ayenew, with a yield advantage of 56.6 per cent, 49.8 per cent and 12 per cent over the local variety respectively. Although Ayenew was appreciated for being early-maturing, its relative susceptibility to bean stem maggot (BSM) made farmers cautious about its potential use. The variability around the average yield also implied that the potential benefit from Dimtu (standard deviation (SD). ±616.6 kg/ha) and Goberasha (SD ±516.4 kg/ha) remained positive up/down compared to the local variety yield (1,237 kg/ha). The local variety was found comparatively inferior both in yield and uniformity. Although Dimtu stood as the top preference after field trials, Goberasha with its comparative yield advantage has also been introduced as an additional option for the producers in the district. Based on the evaluation meeting it was also recognized that Awash Melka and Awash-1 were preferred for the value they have in the market (owing to their colour and shape), and hence were put on the list of potential options for the district.

In Shala district, Nasir was the top preference during the early selection criteria stage (Table 4.4), and exhibited the highest yield advantage (35%) over the local variety, while the other two varieties, Awash-1 and Brown speckled yielded 9 per cent and 23.9 per cent less respectively and were outshone by the local variety. Nasir, in addition to its yield advantage and preferred status based on the initial farmers' criteria, also had relatively good tolerance of moisture stress (which was pronounced early on during that particular season) as well as disease and pest resistance. All together this made it comparatively ideal to the district's farmers. Brown speckled was found susceptible to BSM and other viral diseases during the season. In addition, during the evaluation meeting, farmers expressed that it was difficult to thresh, the stem split at maturity, and it took longer to cook. Though it was picked early on as among the most preferred varieties, its field performance was not good enough to keep farmers' interest once it had cropped. Awash-1, on the other hand, which stood second in farmers' preferences, also gave a lower yield than the established local variety, mainly owing to a low establishment rate early in the season, as a result of moisture stress. During the evaluation meeting, farmers were nevertheless impressed with the yield of Awash-1, owing to it producing the highest number of pods per plant (60–80) and its capacity to yield nearly as much as the local variety, even after an apparently poor stand. *'If we can reap so much with fewer stands, it will be a miracle with better stand population'*, was heard from the farmers, implying that there would be even greater yield potential with a good season and improved management. Farmers had also reported that Awash-1 was in demand in the local market.

As mentioned, the local variety was out-performed in terms of yield only by Nasir. Nasir was also popular in the local market. If cultivated under suitable conditions and with improved management, Nasir could potentially give the highest yield (more than 3,000 kg/ha, mean + SD) and under poor conditions could still give a comparable yield to the local variety (see Table 4.3).

During the evaluation meeting, farmers also showed particular interest in Roba-1, where 182 kg of grain yield was reported by non-trial farmers from 4 kg of seed; Atendaba (110 kg from 4 kg seed); and Dimtu, mainly because of the yield potential and preference to its taste. These varieties stood sixth, fifth and fourth respectively in the early (before the field test) ranking process.

The result of the trial indicated that in Bora district, Tabor had the highest yield advantage (123.8%) over the local, followed by Roba-1 (117.7%) and Awash Melka (16%) (see Table 4.3). Awash Melka had a lower germination rate (compared with the other varieties. Table 4.3 shows that the range within which the potential gains lie (mean ± SD) is more stable for Roba-1 than Tabor, yet the minimum (mean–SD) that farmers may get out of these varieties for Tabor still remains well above the yield of the local variety.

If farmers cannot adequately manage the plot, producing Awash Melka may not offer as good a yield as the established local variety. The FRG farmers in this location showed a distinct preference for common bean varieties with good potential for selling at market (Awash Melka), and for food (Roba-1 and Tabor). In addition to the market value, farmers also attached importance to the early maturing qualities of Awash Melka. Farmers also appreciated the yield potential as well as eating quality of Roba-1 and Tabor.

Qualitative evaluation

In addition to the trial varieties, the FRG farmers in Bora district showed a particular preference for Awash-1 as with Awash Melka, mainly due to its marketability. Other varieties such as Dimtu and Nasir, which were among the lowest-ranked, also captured the farmers' interest, mainly because of their field performance (disease and insect pest resistance and potential yield), as well as taste in local recipes. At the outset, there was almost no interest shown in these varieties (owing to their red colour), but interest rapidly grew when the farmers had the chance to see the performance in the field, and sampled the taste and realized the potential for marketing in neighbouring districts of the southern region.

Table 4.3 Yields of common bean varieties tested at the three sites

Location (N*)	Variety	Days to flowering	Days to maturity	Average yield (kg/ha)	Yield advantage over local		Standard deviation	
					kg/ha	%	kg/ha	As % of the average advantage
Adami Tulu Jido Kombolcha (4)	Ayenew	41	77	1,384.75	147.75	11.90	504.60	342.00
	Dimtu	61	86	1,937.50	700.50	56.60	616.60	88.00
	Goberasha	38	78	1,853.00	616.00	49.80	516.40	84.00
	Local	56	86	1,237.00	0.00		264.00	
Shala (4)	Brown speckled	38	78	1,608.50	(505.50)	(23.90)	876.00	173.00
	Awash-1	38	75	1,924.00	(190.00)	(9.00)	1,098.70	578.00
	Nasir	61	88	2,861.00	747.00	35.30	795.00	106.00
	Local	42	77	2,114.00	0.00		665.00	
Bora (4)	Tabor	42	76	2,915.70	1,612.70	123.80	886.00	55.00
	Roba-1	42	75	2,836.00	1,533.00	117.70	255.70	17.00
	Awash Melka	42	75	1,522.00	219.00	16.80	877.40	401.00
	Local	56	86	1,303.00	0.00		355.60	

* N = number of trial FRG farmers

Table 4.4 Ranking matrix of common bean varieties using qualitative criteria (before field test) in their order of importance, as evaluated by farmers

Location (M/F)*	Evaluation criteria	Nasir	Brown speckled	Awash-1	Atendaba	Mexican-142	Awash Melka	Dimtu	Roba-1	Beshbesh	Zebra	Ayenew	Tabor	Melkie	Goberasha	Gofta
Shala (13/10)	Seed colour	1	2	1	3	1	1	1	1	2	4	1	3	4	2	3
	Cookability	1	1	2	1	3	3	3	2	2	1	3	2	2	4	4
	Taste	1	1	2	1	3	3	3	2	2	1	3	2	2	4	4
	Marketability	1	3	1	3	1	1	1	4	4	4	4	4	4	3	4
	Score	4	7	6	8+	8	8	8	9	10	10	11	11	12	13	15
	Rank	1	3	2	4	5	5	5	6	7	7	8	8	9	10	11
Adami Tulu Jido Kombolcha (15/12)	Seed colour	2	2	1	2	2	1	1	2	3	2	2	3	3	2	2
	Cookability	2	1	3	2	3	3	2	2	2	2	1	2	2	1	1
	Taste	2	2	2	2	2	2	1	1	3	1	1	2	2	1	2
	Marketability	3	2	1	3	2	1	1	3	2	2	2	3	3	2	3
	Seed size	2	1	1	2	2	1	2	2	2	2	1	2	1	1	1
	Score	11	8	8	11	11	8	6	9	12	9	7	12	11	7	9
	Rank	5	3	3	5	5	3	1	4	6	4	2	6	5	2	4

(continued)

Table 4.4 Ranking matrix of common bean varieties using qualitative criteria (before field test) in their order of importance, as evaluated by farmers, (continued)

Location (M/F)*	Evaluation criteria	Nasir	Brown speckled	Awash-1	Atendaba	Mexican-142	Awash Melka	Dimtu	Roba-1	Beshbesh	Zebra	Ayenew	Tabor	Melkie	Goberasha	Gofta
Bora (12/9)	Seed colour	3	3	1	3	2	1	3	3	3	4	3	1	3	2	2
	Cook-ability	2	2	4	3	4	3	2	1	2	2	2	2	2	3	3
	Taste	3	2	3	2	2	1	3	2	2	3	2	1	3	3	3
	Marketability	4	4	2	3	2	1	2	3	3	4	3	3	3	4	4
	Seed size	3	3	2	3	2	1	2	2	3	3	2	3	3	3	3
	Score	15	14	12	14	12	7	12	11	13	16	12	10	14	15	15
	Rank	7	6	4	6	4	1	4	3	5	8	4	2	6	7	7

* Number of male/female farmers of each location involved in the ranking process
1 = best preference; + = comparatively more preferred than others with similar score
Note: Not all the varieties were known at the outset to all the farmers. This made assessing marketability for each variety difficult. Though some of the varieties had qualities which the group appreciated in the early stages, it was difficult for them to say how the market would react. Farmers rated these varieties based on how they felt, given their experience, local markets would potentially respond.

In general, based on the field performance of the tested varieties, in some of the sites, the top varieties selected before field testing also exhibited outstanding field performance. However, this was not the case in all the trial sites. There were certain varieties (not in the top three) which also showed conspicuous field performance and later ranked among the farmers' most preferred (see Tables 4.5 and 4.6). It is, however, important to note that the differences between farmers in terms of land management (plot preparation, proper weeding and so on) and differences in soil fertility between farmers' plots do account for some of the variations in field performance of the same varieties on different plots. With all the criteria combined, the final decisions by farmers on the tested varieties is summarized in Table 4.5.

Based on the preferences showed by the FRG farmers and the field performances of all the varieties, Nasir and Awash-1, in that order, were found to be the best options in Shala district, Dimtu and Goberash in ATJK, and Awash Melka and Awash-1 (for market), and Roba-1 and Tabor (for food) in the Bora district. Other varieties were also taken up as potential options in the respective areas (see Tables 4.5 and 4.6). It seems that the order of importance may change depending on the behaviour of the varieties at market. Nevertheless, fitness for growing in the particular environmental conditions of an area, and field level performance of the varieties remain the most important factors in farmers' preferences.

Enhancing access to seeds of preferred varieties

Following the identification of the best varieties by the respective FRG farmers and the interest created among the surrounding farmers as a result of extension events, arrangements were made to produce seed on farmers' fields, to improve local access to the new varieties.

To respond to the demand created for the best performing varieties, farmer-based seed production was launched in 2006. FRG farmers and district extension officers, including development agents, took part in a training forum, where they learned how to produce and distribute quality seed to the community.

The seed multiplication activity was combined with other on-farm technology (management/cultural practices) evaluation activities. Seeds of the most popular new varieties were grown alongside maize (intercropping) as well as in alternative land preparation and line planting techniques (see Table 4.7). Learning these management practices added useful technical skills and value to the seed multiplication activity by increasing productivity. The trial on various management options, combined with the seed multiplication exercise, served the purpose of both identifying the best management options to enhance productivity and improving local availability of the best seed.

The quantity of seed produced during the 2006 cropping season is indicated in Table 4.8. It should be noted that the figure given for the quantity of seed

Table 4.5 Summary of farmers' assessment of common bean varieties

District	Varieties tested with trial and non-trial farmers	Rank before field test	Field performance			Rank after field test (with remarks)
			Yield	Market	Tolerance to disease and pests	
Shala	Nasir	1st	√√	√√	√√	1st
	Brown speckled	2nd	x	x	x	Dropped
	Awash-1	3rd	√	√√	√	2nd
	Dimtu	5th	√√	√√	√√	3rd
	Roba-1	6th	√√	x	√√	(These varieties were preferred after their field performance owing to their yield and food quality (taste and cook ability)
	Atendaba	4th	√√	x	√√	
Atjk	Dimtu	1st	√√	√	√√	1st
	Goberasha	2nd	√	x	√	Indifferent
	Ayenew	2nd	x	x	x	Dropped
	Awash Melka	3rd	√	√√	√√	2nd (more important at market)
	Awash-1	3rd	√	√√	√√	
Bora	Awash Melka	1st	√	√√	√√	1st for market
	Tabor	2nd	√√	√	√√	Most preferred for food
	Roba-1	3rd	√√	√	√√	
	Awash-1	4th	√	√√	√√	Same as Awash Melka
	Nasir	7th	√√	x	√√	Captured farmers' interest despite rejection after first impression owing to yield performance and tolerance to disease and pests.
	Dimtu	4th	√√	x	√√	

'x' means poor performance; '√√' means best performance

Table 4.6 Summary of the performance and farmers' preference of tested common bean varieties at the three sites

Tested varieties	Performance at the three sites			Remarks
	Shala (kg/ha)	ATJK (kg/ha)	Bora (kg/ha)	
Nasir	2,861**		*	The most preferred variety in Shala. The same variety captured the interest of Bora farmers owing to its yield performance
Awash-1	1,924**	*	**	The number of pods per plant impressed Shala farmers, despite poor population due to moisture stress in the early stages. The same variety was of interest to Bora and ATJK farmers because of demand at market.
Brown speckled	1,608			Susceptible to disease and hard to thresh, therefore not a preferred variety
Dimtu	*	1,937**		Though the market is not yet well developed for this variety, farmers in ATJK favoured it for food and yield performance. Its colour and field performance was also of interest to Shala farmers
Goberasha		1,853*		This variety remains a potential option for ATJK district farmers
Ayenew		1,384		At the outset seen as equally important as Goberasha, but in the field it was susceptible to disease and less popular with the farmers of ATJK
Awash Melka		**	1,522**	Despite lower yield in comparison with others, it was preferred for market. Very poor performance at Shala on non-trial farmers, plots (poor germination)
Roba-1	**		2,836**	High-yielding and preferred for food at Shala
Tabor			2,915**	Yield surprisingly high and liked for food by Bora farmers

** best varietal option for the district; * could be considered as suitable option

produced refers to what the research team and extension workers recorded and does not include the quantity produced from the previous year's harvest of the trial bean varieties.

At the end of the production season, participant farmers were required to return the same amount of seed they were provided for future seed production.

Table 4.7 Seed multiplication exercise (with/without the bean management practices trial)

Seed multiplication	Participant farmers			
	Shala	ATJK	Bora	Total
Combined with trial on cultural/management practice				
With row planting	5	2	4	11
With land preparation using improved plough	5	3	4	12
Without trial				
Sole seed multiplication	2	3	2	7
Total	12	8	10	30

Table 4.8 Farmer-based seed production

District	Variety	Quantity produced (kg)	No. of farmers involved	Remarks
ATJK	Dimtu	1,470	5	
	Awash Melka	2,579	8	
Bora	Awash-1	250	1	Roba-1 produced from seeds of previous harvest not accounted for
	Tabor	250	1	
	Awash Melka	2,579	6	
Shala	Nasir	4,026.6	7	Dimtu and Roba-1 produced from seeds of previous harvest for
	Awash-1	302	1	
	Roba-1	700	1	
Total		12,156.6	30	

The seed thus collected was redistributed to other non-FRG farmers in the respective districts. This 'revolving' of seed from FRG to non-FRG farmers was facilitated with the help of extension workers in the respective district agricultural offices.

The availability of farmer-produced seed for the following season was, however, challenged by two important factors: one, the problem of storability owing to infestation by pests (in Shala), which tempted the farmers to sell the seed as a grain to traders; and two, a lack of proper marketing arrangements for some of the new varieties (in ATJK), which forced some of the farmers to keep and consume relatively large quantities of the seeds produced.

The attempt to improve local availability of the seed was continued in the following years by involving both FRG and a few other non-FRG farmers.

Although the demand for seeds of the preferred varieties was higher than for the locally familiar beans, the local seed production activity only improved the available quantities by a modest level. The experience also paved the way for an organized, farmer-based seed multiplication programme, which is now being advocated by seed companies (for example, the Oromia seed enterprise) and other NGOs, which are trying to create a business opportunity for seed-producing farmers. The activity is felt to have contributed to the improvement of the seed quality that most farmer cooperative unions collect from their members. Nonetheless, this empirical observation may require further investigation.

Challenges and options

The problems observed while working with FRGs at the three sites are listed below in Table 4.9. Even though attempts were made to take counter measures at the time of observing, the challenges are ongoing and probably require continued monitoring and intervention.

Lessons and recommendations

The participatory variety evaluation enabled farmers to understand the range of common bean varieties available to them and to categorize them, using headings such as colour and shape. In addition, the practical training equipped farmers to differentiate different diseases and pests, to recognize the symptoms, and to understand the corresponding management techniques needed to reduce disease incidence. With all such experiences, FRG farmers must be encouraged and given opportunities to share their experience with other farmers, to strengthen farmer-to-farmer learning.

The difference in the performance of the bean varieties in relatively similar physical and socio-economic environments was often attributable to differences in farmers' management practices. There is huge scope for increasing the productivity and quality of any given variety simply by improving management practices. Important consideration should be given to training and improving cultural practices to realize the potential of each variety.

Ensuring active involvement through formal arrangements of the respective district agricultural offices is critical in sustaining the outputs of participatory activity. Despite their busy schedules, it is very important for them to be actively involved in every stage, from planning through evaluation. It may be necessary to have a particular contact liaison person from the respective agricultural offices who can be responsible for sharing information and reporting back to his or her office. Such a flow of information among research, extension and other stakeholders helps in building reliable institutional memory and reduces the impact of staff turnover.

Table 4.9 Challenges observed during the implementation of the participatory variety evaluation process and potential options for better performance

Challenges	Options
In the early stage of the activity, the contact with district agricultural offices was insufficient in that there was a lack of knowledge about the FRG activities being carried out in the respective districts. This checked the smooth implementation of the activity.	It is essential to formalize the linkage through communicating all the details of the activity to the office. It may also be useful to ensure formal engagement of some of the office staff by devoting some of their time to the project. Having a liaison person in the agriculture office can facilitate the flow of information on what activities are being carried out in the district.
The activity was started with an existing FRG which had poor gender composition. Also the physical distance between the group members was great, making follow-up and group management quite complex.	Revising the group composition to make it more representative of the target community, sub-grouping the members with similar socio-economic backgrounds and ensuring they live close to one another, would all help to facilitate follow-up and interaction.
Farmers' understanding of the FRG approach was initially limited to expectations of focus on free gifts/ delivery of inputs (seed and fertilizer) – dependency syndrome.	It is crucial to state clearly, at an early stage, what it means to be a member of an FRG and what duties and responsibilities come with membership. Emphasize knowledge gain and downplay any expectations of free gifts by introducing a cost-sharing mechanism in terms of what the farmer can contribute (size of cost to be shared depends on resource status/capacity of the farmer).
In areas where new varieties showed acceptable agronomic performance, lack of an established market forced farmers to keep more seeds at home for food. This in turn discouraged further seed production.	It is vital to engage extension staff in all activities since they have the widest access to farmers and can facilitate the flow of information throughout the area in the least time, and can create demand using different extension methods. It is also necessary to establish a market information system in the district office and involve the farmers' union in a business plan, setting a clear marketing strategy from the outset, with all relevant stakeholders.
Owing to poor storage facilities, farmers were tempted to sell the seed as grain to traders, which caused problems sustaining the local seed supply.	Develop technologies on improved storage and a centralized local storage facility. Facilitate early sale of seed to farmers so that stock will be spread across many farmers, making storage more manageable for individual farmers keeping relatively small quantities.
Frequent transfer of extension workers (DAs) from one place to another placed stress on the smooth implementation of the activity. There was often a poor exchange of information between incoming and outgoing personnel.	Build institutional memory by strengthening documentation of the activities and processes at all levels – district office and at Farmer Training Centre/*Kebele*. Establish regular information sharing between all stakeholders.

When forming an FRG, maximum care should be taken with regard to the balance and inclusivity of its members, in terms of sex, education, age, wealth status and the physical distance between member farmers to facilitate their accessibility to follow-up visits. However, the criteria also need to be flexible enough to consider the realities in the target sites.

It is essential to ensure from the beginning a shared and clear understanding of the FRG approach to all actors involved. This has a strategic implication for sustainability of the programme. The concepts of cost sharing and capacity development should be stated clearly from the beginning, so as not to foster any notions of dependency.

It is important to develop capacity in researchers as well as extension workers in the essential aspects of the FRG approach, including the management of group dynamics. In addition, designing a mechanism whereby their contribution to FRG activity is recognized is central to motivate extension staff. It can be done through open (public) recognition of outstanding performance in the planning and implementation of agreed FRG-based activities.

In order to secure and sustain stocks of seeds of the preferred varieties through farmer-based seed production, there should be viable storage technology and facilities, as well as training and extension materials on management and control mechanisms of important common bean diseases and insect pests.

Despite the high yield potential of the preferred varieties in the respective districts, the absence of market information is a constraint to continued production. There needs to be, from the outset, a mechanism to create wider demand and to provide market information for farmers, either through the district office or farmer associations.

Conclusion

Farmer preference for a variety does not always depend on yield alone. There are other important factors, such as colour, size, cooking time required and state after being cooked, taste, market demand, maturing period, and resistance to disease and pests. It is also important to note that the emphasis on these different qualities changes from one district to another. Accordingly, the favourite varieties in Shala were Nasir, Awash-1, Roba-1 and Dimtu. In Adami Tulu Jido Kombolcha the favourites were Dimtu, Awash Melka, Awash-1 and Goberasha. Awash Melka, Awash-1, Roba-1 and Tabor (in that order) were farmers' preferences in Bora district for either food or market. There were still other varieties which appealed to farmers and could be candidates for future production.

On-farm participatory technology evaluation and promotion with well-organized, properly facilitated farmer groups working together with researchers and extension workers proved a valuable approach for rapid technology evaluation and dissemination. Research and extension activities need to capitalize on the FRG approach, as it accommodates well to differences in local socio-economic circumstances, and offers a reflective learning environment for all actors involved. The FRG approach also creates demand for research and institutions to focus on farmer-preferred traits and to work on making preferred technologies more widely accessible.

About the authors

Endeshaw Habte is a researcher and coordinator of the socio-economics programme at Melkassa Agricultural Research Centre. He holds an MSc in agricultural extension from India and currently is studying for a PhD in agricultural economics at Haramaya University, Ethiopia.

Kidane Tumsa has been working both as breeder and pathologist in the bean improvement programme at Melkassa Agricultural Research Centre since 2003. He obtained his MSc at Haramaya University, Ethiopia, in 2007.

Dr Berhanu Amsalu Fenta is a breeder in lowland pulse crops, including the common bean. He coordinates the national pulse, oil and fibre crops research programme of the Ethiopian Institute of Agricultural Research.

Mr Abiy Tilahun is an entomologist working on beans and other crops at Melkassa Agricultural Research Centre. He obtained his MSc in agricultural entomology at Haramaya University, Ethiopia, in 2004.

References

Abate, T. (1995) 'Pest management in lowland pulses: progress and prospects', in H. Assefa (ed.) *Proceedings of the 25th Anniversary of Nazret Agricultural Research Center: 25 Years of Experience in Lowlands Crops Research*, 20-23 September 1995, pp.181-194, Nazret Agricultural Research Center, Nazret, Ethiopia.

Asefa, T., Reda, F. and Amsalu, B. (2006) 'Promotion of improved haricot bean production systems in East Shewa: An example of a successful partnership among stakeholders', in T. Abate, (ed.), *Success in value chain: proceeding of scaling up and scaling out agricultural technologies in Ethiopia: An International Conference*, 9-11 May 2006, pp. 21-31. EIAR, Addis Ababa, Ethiopia.

Ellis, F. (1993) *Peasant economics: farm households in agrarian development*, Cambridge, Cambridge University Press.

Food and Agriculture Organization (FAO) (2007) 'Scaling up of good practices: policy brief 21', in *Sustainable Agriculture and Rural Development (SARD) Initiative* [website] < www.fao.org/sard/initiative> [accessed 6 January 2014].

Feder, G., Just, E. R. and Zilberman, D. (1985) 'Adoption of agricultural innovations in developing countries: a survey', *Economic Development and Cultural Change*, 33, pp. 255-98.

Mellis, D. and Mwaniki, B. (1999) 'Participatory technology development for animal traction: experience from semi-arid areas of Kenya', in P. Stakey and P. Kaumbutho (eds.) *Meeting the Challenges of Animal Traction. A Resource Book of the Animal Traction Network for Eastern and Southern Africa* (ATNESA), Harare, Zimbabwe, pp. 20-27, London, Intermediate Technology Publications.

Nigatu, D., Girma, T. and Abebe, A. (1995). 'Lowland pulse improvement in Ethiopia', in H. Assefa (ed.) *Proceedings of the 25th Anniversary of Nazret Agricultural Research Center: 25 Years of Experience in Lowlands Crops Research, 20-23 September 1995*, pp. 40-47, Nazret Agricultural Research Center, Nazret, Ethiopia.

Ponniah, A., Puskur R., Werkneh, S. and Hoekstra, D. (2008). *Concepts and Practices in Agricultural Extension in Developing Countries: A Source Book.* International Food and Policy Research Institute, Washington, DC, USA, and International Livestock Research Institute, Nairobi, Kenya.

Willem, H. and Bertus, W. (2004) *Building Social Capital for Agricultural Innovation: Experience with Farmer Groups in Sub-Saharan Africa*, Royal Tropical Institute (KIT), Amsterdam, Netherlands.

Yetneberk S. (1995) 'Quality evaluation of lowland crops: progress and future direction', in H. Assefa (ed.) *Proceedings of the 25th Anniversary of Nazret agricultural Research Center: 25 years of experience in lowlands crops research, 20-23 September 1995*, pp.259-267, Nazret, Ethiopia.

CHAPTER 5

Lowering teff seeding rate using a seed spreader via the participatory approach in South Ethiopia

Fanuel Laekemariam, Gifole Gidago and Wondemeneh Taye

Abstract

Participatory research on teff seeding rates using a seed spreader in Wolaita Zone, South Ethiopia, was conducted during 2010–12. Six seeding rates (5, 10, 15, 20, 30 and 35 kg/ha) were investigated in 2010–11, and in 2012 rates of 5–15 kg/ha were evaluated. Rates lower than 20 kg/ha were mixed with soil. The seeding rate significantly influenced plant height, pannicle length, tillers per plant, days to maturity and lodging, whereas straw and grain yield were not significantly influenced. Farmers preferred rates of 5, 10, 15 and 20 kg/ha in that order. They commented on the lesser crop establishment on 5 kg/ha at times of heavy rain and increasing seedling establishment for 15 kg/ha and above. Considering all the relevant data and economic aspects, 10 kg/ha was found to be the safest rate. The range between 5 and 10 kg/ha is also feasible under farmers' conditions. Further investigation on teff varieties, fertilizer rates and sowing methods is suggested.

Keywords: Ethiopia, Farmer Research Groups, participatory research, seeding rate, spreader, teff

Background and justification

Teff (*Eragrostis tef*) is a cereal indigenous to Ethiopia which is highly valued by both farmers and consumers (Hailu et al. 2001). It is also a leading crop in terms of area coverage (CSA, 2013). Farmers' seeding rate for teff varied between 20 kilograms per hectare (kg/ha) in Sendafa Aleltu to 120 kg/ha in the Adet area (Kenea et al. 2001), while existing research recommends seeding rates of 25–30 kg/ha (ESE, 2001; Kenea et al. 2001). Difficulties in even and uniform distribution when sown by hand, soil type, tillering capacity, uncertainty of germination and weed suppression at early stages of growth were all reported as factors in farmers choosing higher seeding rates (Fufa et al. 2001; Kenea et al., 2001; Tefera and Belay, 2006). These days, teff production for grain and/ or fodder purposes outside Ethiopia in such countries as the USA, South Africa and Australia has been reported (Twidwell et al. 1991; Stallknecht et al. 1993; Lacey and Llewellyn, 2005; Hall and Cherney, 2010). They have accomplished sowing using planters with a seeding rate of 4.5–9 kg/ha and reported grain yields of 0.7–1.4 tonnes per hectare (Twidwell et al. 1991) and 2 t/ha (Lacey and Llewellyn, 2005).

The average national yield of teff in Ethiopia is 1.4 t/ha (CSA, 2013). However, yields of 1.7–2.5 t/ha are easily attainable using improved varieties and recommended management practices (Hailu and Seyfu, 2001). In addition, yield potential from 3.4 to 5.0 t/ha has been reported (Hailu and Seyfu, 2001). In Wolaita, South Ethiopia, where teff is an important grain crop, the average productivity still remains as low as 0.7 t/ha (SNNPR, 2007).

Teff grains are the smallest in the world, taking 150 grains to weigh as much as one grain of wheat, and there are 2,500–3,000 seeds to the gram (Lacey and Llewellyn, 2005). Its 1,000-seed weight is about 0.3–0.4 grams (Stallknecht et al. 1993). Among the agronomic constraints in Ethiopia, the tiny seed size of teff could be affecting the attainment of even and uniform seed distribution when sown by hand (Seyfu, 1997; Fufa et al. 2001). This in turn affects optimum plant population, which in turn affects plant competition, lodging (flattening of standing plants), and disease and insect prevalence, all of which could contribute to lower productivity.

Seed spreaders (mixers) ease the hard work of seed sowing and assist the even distribution of seed. Spreaders include dry and fine materials like sand, soil and compost. A number of reports from different countries have shown the importance of seed spreaders in avoiding the limitations imposed by the tiny seeds. Rathore (2001) reported the experience of Indian farmers in mixing sesame seed with dry sand. Similarly, Naturland (2002) reported the trend of mixing sesame seeds with sand, soil, ash or dried and sieved manure or compost. Owuor et al. (2001) reported the experiences of Kenyan farmers using sand seed spreaders for finger millet (*Eleucine spp.*). As these authors have stated, seed spreaders help to create bulk, reducing the amount of seed required. Spreaders also aid the uniform broadcasting of tiny seeds by hand, uniform growth and potentially better yields. In view of this, seed spreaders seem highly suitable for use with teff to solve the limitations and consequences associated with its size. In addition, using spreaders could save a considerable amount of seed at both household and national level, creating an opportunity for breeders to disseminate improved teff varieties, and thus contribute to attaining greater food security.

So far, very few trials have been reported where sand is used as a seed spreader. Mulu et al. (1994–5, cited in Fufa et al. 2001) concluded that the use of a spreader (other than sand) in combination with seeding rates of more than 25 kg/ha was ideal at Debre Zeit and Akaki, whereas 10 kg/ha mixed with sand with soil compaction at Melkassa, and 10 and 20 kg/ha at Boset district proved to be optimal (FRG extension material series, n.d.). However, in Ethiopia generally, including Wolaita, information is lacking regarding the use of a seed spreader for lowering the teff seeding rate. Research was therefore commissioned to test the following hypothesis: 'Mixing teff seeds with soil under existing farmer practices and participation will reasonably reduce the seeding rate without compromising the grain yield and economic returns'. To test this, participatory research with farmers under their current practices was conducted with the following specific objectives:

- to evaluate the growth and productivity of teff seeding rates using a seed spreader;
- to identify the most suitable seeding rate for the study area;
- to facilitate knowledge transfer and enhance farmers' participation in research activities.

Research methodology

Description of the study area

This experiment was conducted in South Ethiopia in Wolaita Zone, Duguna Fango district (Edo *kebele*[1]), in the main rainy cropping season of 2010 and 2011. The research site is found at an altitude of 1591 metres above sea level (latitude 07°02'14.9" N and longitude 038°00'44.5" E), where the minimum average annual temperature is 16°C and maximum 26°C, and the average annual rainfall is 950 mm. The soil of the study area is sandy loam with a neutral soil pH. The organic carbon content is very low. The available phosphorous (P) and total nitrogen (N) content was medium and very low respectively. The site was selected because teff is intensively cultivated in the area and also due to the presence of an established Farmer Research Group.

Multidisciplinary team collaboration and interaction

The research was based on a multidisciplinary and multi stakeholder team approach where the team comprised researchers from Wolaita Sodo University, the Farmer Research Group (FRG), and development agents (DAs)[2]. The university research team was composed of agronomy, soil and crop protection experts. FRG members were first established by DAs for research and extension purposes in 2009. There were 10 members (7 male and 3 female) during 2010 and 16 (12 male and 4 female) during 2011 and 2012. The FRG members had experience of participatory research on integrated nutrient management in the area. The established FRG was ideal, considering their research and extension experience in their communities.

Research design

Bio physical research design. Six teff seeding rates – 5, 10, 15, 20, 30 (recommended rate), and 35 kg/ha⁻ (average farmers' practice) – were evaluated. The researchers originally designed the trial to test teff seeding rates up to 30 kg/ha; however, before starting the experiment in the first year, participating farmers mentioned that due to the small grain size of teff, they were accustomed to using an average 35 kg/ha seeding rate. In view of that, the research stakeholders (researchers, FRG members and DAs) agreed to include 35 kg/ha as the existing local seeding practice.

The seed rates of 5, 10, 15 and 20 kg/ha were mixed with dry sand (in 2010) and soil (in 2011). The spreader was collected from farmers' fields. As extraneous material such as small sticks and pebbles affects the applicability of the seed spreader, coarse materials and plant debris were picked up and sun-dried prior to sieving. We used locally made sieves (used for small grain cereals). Sieving produces soil particles that are slightly larger than the teff grains so that seed mixes easily, uniformly and handles comfortably when being sown.

The experiment was conducted as a randomized complete block design with four replications at one trial field. Seeding rates in the first block were arranged in an easy-to-grasp way for the farmers according to the increasing order of the seeding rate (i.e. 5 to 35 kg/ha). Seeding rates on the remaining replications were assigned using a randomization technique. The size of each plot was 9 square metres (3 m x 3 m). The spaces between plots and blocks were 0.75 m and a 1 m free area was left around the trial field.

All the preparation activities were completed via participatory processes. The seed and spreader preparation procedure consisted of the following steps. First, we prepared dried soil, locally made sieves, empty buckets and plastic/polyethylene bags/sheets. In the second step, dried soil was sieved and collected in a bucket. In the third step we calibrated (determined the volume) of farmers' average seeding rate (35 kg/ha). For this step, the farmers' seeding rate was placed in an empty container and its volume was determined and/or marked. The mark was then used to indicate what level of seed spreader should be added on the top of lower seed rates in order to be consistent. In step four, the 35 kg/ha containing seed bucket was emptied. The lower seed rate (e.g. 10 kg/ha) was measured and put it in a calibrated empty bucket. The space between the teff seed and marked level was observed. Then, we added sieved soil on top of the seed up to the level mark. Note that, regardless of the differences in the seed amount (i.e. 35 kg/ha or 10 kg/ha), the total volume of seeding materials would be the same. By doing this, the feeling of holding the volume of seed and soil together would be familiar to the farmers' usual experience of sowing. Step five consists of seed and soil mixing. In this process, seed and soil was placed on a plastic sheet and mixed thoroughly. This mixture was then used for sowing.

Design of participatory procedure. Farmers were full partners in the experiment, including initial discussions on cost sharing and participatory evaluation. This made the participatory research more effective than a conventional approach, owing to responsibility sharing, smaller direct costs being borne by the university, and easier information dissemination by and between farmers (via the extension services). During our discussions, researchers, FRG members and DAs reached common understandings on the following points about the contribution of farmers and DAs to the research.

- Farmers would share costs on ploughing, sowing, labour and the provision of locally available materials, including marker pegs/labels, whereas researchers were responsible for making available the seed and fertilizer.
- Farmers would provide the land and the output (grain and straw) would be given to the owner of the trial field.
- FRG members would share responsibility for careful follow-up and guarding the research field from theft, animals, etc.
- Information would be disseminated (shared) to non-participant farmers (through extension services).
- Since the landholding size per household is very small (less than 0.5 ha), farmers would receive a minimum compensation fee for the empty spaces (between plots and blocks).
- DAs would coordinate farmers, information sharing with other farmers, field follow-up and reporting weather conditions, etc.

In addition, initial hands-on training was given to participant farmers on how seed is mixed with seed spreaders and the purpose of mixing, and crop management practices were explained. Researchers took responsibility for agronomic data measurements, soil sample collections and so on, keeping farmers briefed to help their overall understanding of the experiment. Farmers contributed various ideas during the experiments which were used to improve the quality of our research.

Crop management

Teff variety DZ-Cr-37 was used for the experiment. It is early maturing and was the preferred choice of participating farmers. Land was prepared using oxen ploughs. It was ploughed four times before planting and the last ploughing was used for sowing. The seedbeds were levelled and sowing was done by hand broadcasting. Fertilizer was applied in the form of urea (46-0-0) at a rate of 50 kg/ha and di-ammonium phosphate (DAP) (18-46-0) at 125 kg/ha. DAP was applied at sowing time while urea was applied by split application (half at planting and the remaining half at mid-tillering stage). Fertilizer applications and weeding were uniformly carried out on all experimental plots as often as required. Disease and insect prevalence was checked regularly.

Data collection and analysis

Agronomic data. Data such as plant height, pannicle length, average number of tillers per plant and number of fertile tillers per plant were collected. The random sampling method was applied. For data measurement, 10 randomly selected plants were manually uprooted at physiological maturity stage. Plant height was measured from the base of a mature plant to the tip of the pannicle. Pannicle length was measured from the node where the primary pannicle branch starts to the tip of the plant. Number of tillers/plant was the count of tillers which arose from a single plant and number of fertile tillers/plant was the count of tillers which had fertile pannicles.

Days to physiological maturity and lodging percentage were also recorded. Days to physiological maturity was visually determined in terms of days from emergence until the attainment of physiological maturity of 75 per cent of the stand in each plot. This was indicated by the senescence of the leaves as well as free-threshing of grain from the glumes when pressed between the forefinger and thumb. Lodging was assessed close to maturity stage by looking at the degree of inclination of plants. Biological and straw yield, grain yield, and harvest index data were also collected after harvesting the whole plot at ground level, using sickles. For the biological yield, the whole plants were sun dried until ready for manual threshing and then measured. After threshing, the grains were separated to measure the grain yield. The straw yield was measured by subtracting the grain yield from the above-ground biomass yield. The harvest index is expressed in percentage terms as the ratio of grain yield to biological yield.

Farmers' evaluation procedure. As the seed/soil mixing approach was new to farmers, they were very closely interested in the experiment and discussed the results with the research team at each visit to the site. Participant farmers always observe and judge crop growth at different stages. In our discussions we were able to identify that farmers were not only looking at the yield but also emergence, density, plant colour, stem thickness and lodging. Accordingly, we took notes of farmers' opinions and observations at the early stage of plant development, the vegetative stage (prior to pannicle initiation) and at crop stand evaluation prior to the crop harvest.

A field day was organized as the crop was reaching maturity to evaluate farmers' preferences. Farmers were placed into four groups and each one was assigned to one set of treatments. Farmers were encouraged to use their knowledge to evaluate the plant stand, including tillering, pannicle length, observed seed set on pannicle, and expected straw and grain yield. Each group's preferences were recorded by a secretary. The group evaluated all the treatments on a block until common agreement was reached and all preferences were ranked. In the course of this, there was intense discussion around whether each seeding rate was performing well or not in relation to the evaluation criteria. Finally, the group reached consensus and came up with common ranking preferences to be presented to other groups. All their discussion points and comments were collected. To summarize the farmers' ranking order, the tally method was used in which the first, second and third ranking had a weighted value of three, two and one points respectively.

Partial economic analysis. The following steps were used for partial budget analysis:
1. Total income = Gross income (birr) = income from grain yield + straw
2. Total variable costs (birr) were seed and fertilizer costs
3. Net benefit (birr) = gross return– total variable cost
4. Benefit to cost ratio = ratio of net benefit to total variable costs

Experiment on farmers' preferred seeding rates. For two years (2010 and 2011), six seeding rates were evaluated. After that, farmers were consistently interested in using 5 kg, 10 kg and 15 kg seeding rates and suggested their need to evaluate on larger plots (6 m x 6 m) in 2012, rather than doing it all again on small plots. The trial was carried out on one farmer's field, using teff variety DZ-Cr-37. The spreader preparation process was similar to 2010 and 2011. The seeding rates were evaluated at three replications. Crop management and data collection procedures were the same in 2012 as in 2010.

Statistical analysis. Data obtained from the experiment were analysed using the general linear model of the Statistical Analysis System software (SAS, 1997). Effects were considered significant if p values were less than 0.05. Differences between treatment means were tested using the least significant difference (LSD) test at 5 per cent level of significance. Economic analysis calculation was performed using Microsoft Excel sheets.

Results and discussion

Crop development and growth parameters

Teff seeding rates during the 2010 and 2011 cropping seasons exhibited significant variation. The longest (90.9 cm) and shortest (83.2 cm) plant height during 2010 was recorded from 10 kg/ha and 30 kg/ha seeding rates respectively, whereas during 2011, the tallest (87.51 cm) and shortest (77.71 cm) plant height was recorded from 15 kg/ha and 35 kg/ha seeding rates respectively (see Table 5.1). In general, increasing seeding rates significantly decreased plant height. As indicated by Anwar et al. (2011), taller plants at lower seeding rates conflict with general perceptions. The reason might be that the relatively uniform seed distribution and less competition between teff plants taking nutrients at lower seeding rates allows the plants to attain more height when compared with higher seeding rates. Contrary to our findings, Gobeze et al. (2007) reported non-significant differences in the plant height of triticale at different seeding rates.

The number of days taken to reach physiological maturity was measured during 2010, with seeding rates resulting in significant differences. Higher seeding rates (30 kg/ha and 35 kg/ha) required 60 days to get to maturity stage, whereas 5 kg/ha required 71.5 days to get to maturity stage. There was, however, no significant difference with 10 kg/ha (see Table 5.1). The seeding rates 30 kg/ha and 35 kg/ha reached physiological maturity 19 days earlier than 5 kg/ha. Generally, physiological maturity was enhanced as the seeding rate increased. Lower intra-specific competition for resources might result in delayed maturity. In contrast to our findings, Anwar et al. (2011) on rice reported delayed maturity at higher seeding rates compared with lower seeding rates.

Seeding rates did not exert significant influence on dry biomass and straw yields. However, during 2010, the highest dry biomass and dry straw yields came from 35 kg/ha, while the lowest dry biomass and dry straw yields were recorded from 5 kg/ha (see Table 5.1). Similarly, in 2011 the highest and lowest dry biomass and straw yields were recorded from 35 kg/ha and 5 kg/ha seeding rates respectively. Even so, the seeding rate influence was insignificant; numerical values showed an increasing trend from low to higher seeding rates. The reason for the recorded non-significant differences might be because the highest number of tillers per plant and the tallest plant height in the lower seeding rates compensated for higher biological and straw yield of standard check (30 kg/ha) and farmers' practice (35 kg/ha). Similarly, Gobeze Loha et al. (2007) reported non-significant differences with triticale in biomass yield between varied seeding rates.

The seeding rate exhibited a significant influence on the lodging percentage of teff (see Table 5.11). During 2010, the highest (75%) and lowest (10%) percentage lodging were recorded from 35 kg/ha and 5 kg/ha seeding rates respectively. Similarly, during 2011 the highest (63.05%) and lowest (19.65%) lodging percentages were recorded from 35 kg/ha and 5 kg/ha seeding rates respectively. We realized that as the seeding rate increased, there was an increasing lodging percentage trend. Our findings were in agreement with Fufa et al. (2001), who reported a reduced lodging percentage rate at lower plant populations of teff. Moreover, Yenesew (n.d.) also reported that reducing the teff seeding rate from 25 to 10 kg/ha minimized lodging problems.

Yield components and grain yield

Varying seeding rates has resulted in significant variations on pannicle length and total number of tillers and fertile tillers per plant (Table 5.2). Data regarding panicle length showed a significant variation depending on seeding rates. The longest (34.1 cm) and shortest (30.1 cm) pannicle length recorded during 2010 were from 10 kg/ha and 35 kg/ha respectively. The longest (32.14 cm) and shortest (27.65 cm) pannicle length during 2011 were recorded from 20 kg/ha and 35 kg/ha respectively. The findings were in agreement with Anwar et al. (2011), who reported a significant influence and decreased pannicle

Table 5.1 Development and growth characteristics of teff for different seeding rates mixed with a spreader, 2010 and 2011

Seeding rate (kg/ha)	Plant height (cm)	Physiological maturity (DAE)	Biological yield (t/ha)	Straw (t/ha)	Lodging (%)
		2010			
5	88.35ab	71.50a	3.71	2.86	10.00c
10	90.90a	70.75a	4.18	3.16	17.75c
15	88.98ab	67.50b	4.29	3.24	26.25bc
20	85.93bc	63.75c	4.39	3.34	36.75b
30	83.15c	60.00d	4.80	3.64	67.00a
35	85.43bc	60.00d	4.83	3.69	75.00a
LSD $_{0.05}$	4.50	2.66	ns	ns	17.24
CV (%)	3.43	2.69	11.47	11.13	29.50
		2011			
5	85.45a	-	3.31	2.43	19.65d
10	84.78a	-	3.50	2.60	26.75cd
15	87.51a	-	4.06	3.05	34.95c
20	85.45a	-	4.07	3.04	46.25b
30	78.83b	-	3.92	3.00	53.80b
35	77.71b	-	4.23	3.23	63.05a
LSD $_{0.05}$	5.64	-	ns	ns	9.23
CV (%)	4.50	-	13.40	13.74	15.03

DAE = days after emergence, lSD = least significant difference, ns = no significant difference, CV = coefficient of variation

Note: values assigned with the same letters are not significantly different from each other.

length at higher seeding rates. In addition, significant effects on spike length of triticale due to varied seeding rates were reported by Gobeze et al. (2007).

The highest total number of tillers per plant (5.9) and fertile tillers per plant (4.83) in 2010 were obtained at 10 kg/ha, whereas the farmers' seeding rate (35 kg/ha) gave the smallest total number of tillers per plant (2.88) and fertile tillers per plant (2.45). The 10 kg/ha seeding rate showed a 74 per cent and 67 per cent higher total number of tillers per plant and number of fertile tillers per plant respectively, over the standard check (30 kg/ha). Furthermore, the 10 kg/ha seeding rate had a higher total number of tillers per plant (105%) and higher number of fertile tillers per plant (97%) over farmers' average seeding rate (35 kg/ha). Likewise, during 2011, the maximum tillers and fertile tillers per plant were recorded from 10 kg/ha, whereas the minimum for both parameters were recorded from the 30 kg/ha seeding rate. Our findings revealed that lower seeding rates resulted in higher tillers and fertile tillers per

Table 5.2 Grain yield, yield attributing parameters and harvest index of teff for different seeding rates mixed with a spreader, 2010 and 2011

Seeding rate (kg/ha)	Grain yield	Panicle length	Total no. of tillers/plant	No. of fertile tillers/plant	Harvest index
	(t/ha)	(cm)			(%)
2010					
5	0.85	33.18ab	5.45a	4.58ab	22.68
10	1.02	34.08a	5.90a	4.83a	24.50
15	1.05	33.23ab	4.98ab	3.88bc	24.55
20	1.05	32.80ab	4.05 bc	3.43cd	23.88
30	1.16	31.35bc	3.40cd	2.90de	24.13
35	1.14	30.075c	2.88d	2.45e	23.53
$LSD_{0.05}$	ns	2.23	1.02	0.88	ns
CV (%)	15.97	4.56	15.30	15.85	8.89
2011					
5	0.88	31.56a	4.47a	3.57a	27.25a
10	0.90	31.53a	4.54a	3.45a	25.75ab
15	1.01	31.50a	4.14a	3.12ab	25.00abc
20	1.04	32.14a	3.24b	2.6bc	25.50abc
30	0.92	28.61b	2.73b	2.10c	23.25c
35	0.99	27.65b	2.78b	2.41bc	23.75bc
$LSD_{0.05}$	ns	2.27	0.78	0.72	0.023
CV (%)	14.20	4.94	14.11	16.61	6.10

LSD = least significant difference, ns = no significant difference, CV = coefficient of variation
Note: values assigned with the same letters are not significantly different from each other.

plant compared to higher seeding rates. In general, decreasing the seeding rate led to an increase in tillers per plant. This might be due to lesser competition of plants for nutrients which ultimately leads to increased plant growth. A similar result was also reported by Gadalla (2010), who found that lower seeding rates of Rhodes grass resulted in increased tiller numbers per plant.

Grain yield. Grain yield is a function of the integrated effects of yield components, which are in turn influenced by growing conditions. In teff, the significant functions are pannicle length, tillers per plant, and number of plants per unit area. The analysis of results during 2010–11 showed seeding rates as statistically non-significant influences on the grain yield of teff. However, the numerical values during 2010 showed an increasing trend up to 30 kg/ha and then decrease. The value in 2011 showed an increasing trend up to 20 kg/ha and thereafter a decline. The highest (1.16 t/ha) and lowest (0.85 t/ha) grain yield during 2010 were noted from 30 kg/ha and 5 kg/ha seeding rates respectively. During 2011, the highest (1.04 t/ha) and the lowest (0.88 t/ha) yields were recorded from 20 kg/ha and 5 kg/ha seeding rates respectively. The non-significant differences among seeding rates on the grain yield of teff could be because of the compensatory effects of yield components such as panicle length, tillering and plant density. In agreement with our findings, Fufa et al. (2001) came up with the same result for the seeding rates for teff when varied between 10 and 40 kg/ha, due to the compensatory ability of the teff plant through increased tillering. In addition, the superior performance of 10 kg/ha mixed with sand on compacted soil over 20 and 30 kg/ha seeding rates at Melkassa area, and the more or less equivalent performance between 10 and 20 kg/ha at Boset district, were reported in the FRG extension material series (n.d.). Yenesew (n.d.) also stated a potential yield rise to 5/ha by reducing the teff seeding rate from 25 to 10 kg/ha. Furthermore, a factorial combination of four seed rates (15, 25, 40 and 55 kg/ha) and three filler ratios (0:1, 10:1 and 20:1 sand to teff) on the grain yield of teff (variety DZ-01-354) was tested at Debre Zeit (both on Vertisol and Inceptisol, which are different types of soil in terms of size of particles and water-holding ability) and Akaki (Vertisol) (Mulu et al. 1994–5, cited in Fufa et al. 2001). In agreement with the present findings, the grain yield was not significant at all three locations. The authors concluded that the use of a filler (other than sand) in combination with seeding rates of over 25 kg/ha was a practical approach to alleviate the problem of attaining even and uniform seed distribution when sown by hand broadcasting.

Harvest index. The relationship between grain yield to total biological yield of crop is expressed in terms of the 'harvest index', which determines the capacity to convert dry matter into grain yield. Statistical analysis results showed non-significance during 2010 but significance in 2011. The highest (24.63%) and lowest (22.63%) harvest indices were recorded from 15 kg/ha and 5 kg/ha respectively (see Table 5.2). During 2011, the maximum (27.25%) and the minimum (23.25%) were obtained on 5 kg/ha and 30 kg/ha seeding rates respectively. From this analysis, it may not be clear whether the physiological ability of teff plants to utilize dry matter towards an economic yield are affected to any significant extent by seeding rate.

Farmers' opinion and evaluation results

The FRGs consisting of 10 members (seven male and three female) during 2010 and 16 members (12 male and four female) during 2011 and 2012 were mainly involved in field evaluation. As the seed/soil mixing approach was new to them, farmers were initially suspicious of the experiment. When we visited the site at different times, we had discussions with farmers whom we identified as being open to the participatory approach. We were potentially asking farmers not just to observe and judge the growth of the crop at different stages, as they usually do. We also wanted them to observe emergence, density, plant colour, stem thickness and lodging. Unfortunately, early on in the experiment there was heavy rain in the first week of seed sowing (before crop emergence). As the seeds are so tiny, they are left at a very shallow depth, virtually on the surface. This left the seed exposed to the rain and the seed was either washed away or buried deeper – a common worry of the participating farmers using the 5 kg/ha seeding rates. Another worry was the moderate lodging on 15 kg/ha and higher lodging particularly on 20 kg/ha and higher seeding rates, which was associated with dense plant stands.

Farmers' observations prior to pannicle formation were concerned with differences in colour and stem thickness. They noted that there was a dark green colour and thicker stems on plants with lower seeding rates and a yellowish colour and thinner stems as the seed rate was increased.

Table 5.3 Summary of farmers' preferences during crop stand evaluation, 2010 and 2011

Seeding rate						
5	6	4	5	6	21	1st
10	5	5	6	4	20	2nd
15	3	6	3	5	17	3rd
20	4	3	4	3	14	4th
30	3	3	3	3	12	5th
35	3	3	3	3	12	6th
2011						
5	2	6	6	6	20	1st
10	5	5	5	3	18	2nd
15	6	3	4	4	17	3rd
20	3	4	3	5	15	4th
30	4	2	1	2	9	5th
35	1	1	2	1	5	6th

Note: Higher points indicate a higher rank, so 6 points indicates 1st rank in that group, and so on.

Crop stand evaluation was performed prior to crop harvest. The farmers' evaluations were made by ranking. Data suggests that participant farmers paid more attention to the lower seeding rates. During the 2010 cropping season, 5 kg/ha, 10 kg/ha, 15 kg/ha and 20 kg/ha were preferred as the first, second, third and fourth rank respectively (see Table 5.3). A similar trend of preference ranking was noticed during the 2011 cropping season. Visual observation of pannicle length, lodging intensity, expected grain and straw yield were the most frequently indicated justifications for farmers ranking the treatments as they did. They further indicated that teff plots with lower exposure to lodging (i.e. lower seeding rates) were preferred because they were less prone to shattering during heavy rain. In addition, farmers said that the grain and straw yield from the lower seeding rates were on a par with the grain and straw yield from higher seeding rates.

For two years (2010 and 2011) six seeding rates were evaluated. After the trial, farmers were consistently interested in 5 kg/ha 10 kg/ha and 15 kg/ha seeding rates and suggested their need to evaluate on a wider scale (6 m x 6 m plots) in the year of 2012. The trial was carried out on one field. The seed, spreader preparation and processes were all similar to 2010 and 2011. Three replications of each seeding rate were evaluated. Growth and yield attributing parameters and the grain yields of the farmers' preferred lower seeding rates showed statistically non-significant differences (see Table 5.4). Partial economic analysis during the 2012 cropping season indicated greater benefits per unit cost applied on 5 kg/ha followed by 10 and 15 kg/ha seed rates (see Table 5.5). The farmers' crop stand preference indicates that 10 kg/ha, 5 kg/ha and 15 kg/ha were ranked first, second and third respectively (see Table 5.6).

In order to draw conclusions on the proper seeding rate, we used average performances across three years (2010 to 2012). Grain yield was similar on 10 and 15 kg/ha but a little lower on 5 kg/ha (see Figure 5.1). The straw yield increased with the increase in seeding rates. The lower grain and straw yield on 5 kg/ha in our participatory research might be associated with the seeds' vulnerability at sowing time. Benefit to cost ratio (2010–12) indicates more or less similar performances between seeding rates (see Table 5.7). The farmers' evaluation during crop maturity stage (2010–12) indicated that their preferences were between 5 to 10 kg/ha, and 15 kg/ha was the least preferred rate (see Table 5.8).

Our research findings throughout the three years of evaluation were that the 10 kg/ha seeding rate was the safest for farmers in terms of its technical and economic feasibility, consistent distribution while sowing, better coverage and less risk during plant emergence under heavy/erratic rain or dry conditions. Our suggested seed rate is comparable with the finding of Stallknecht et.al. (1993) in the USA, who indicated 5–8 kg/ha teff seed as the optimum rate using machine planters.

The contribution of farmers to improving research quality

Farmers' participation is helpful in order to establish precise information during the research period and helping to make the research findings applicable in a wider context. The importance of farmers' participation in research at all stages was reported by Abule et.al. (2011) and Nishikawa (2011). In this trial at Edo kebele, farmers played a vital role in the following:

- shifting the idea of using sand as a seed spreader (2010) to soil (2011 and later), as access to soil is easier;

Table 5.4 Growth and yield attributing parameters and grain yield of teff at lower seeding rates, 2012

Seeding rate (kg/ha)	Plant height (cm)	Panicle length (cm)	Total tillers/plant (no.)	Fertile tillers/plant (no.)	Biological yield (t/ha)	Straw yield (t/ha)	Grain yield (t/ha)	Harvest index (%)
5	99.53	33.50	4.73	3.57	4.95	3.61	1.34	27.0
10	99.87	32.50	4.10	3.13	5.49	4.18	1.31	24.3
15	97.20	34.30	5.10	3.70	5.38	4.10	1.28	24.5
LSD $_{0.05}$	ns	ns	ns	ns	ns	ns	ns	ns
CV (%)	3.50	3.80	11.15	6.70	16.60	23.90	13.90	23.50

LSD = least significant difference, ns = no significant difference, CV = coefficient of variation

Table 5.5 Partial economic analysis of grain and straw yield, 2012

Parameters	5 kg/ha	10 kg/ha	15 kg/ha
Grain yield (kg/ha)	1,340	1,310	1280
Grain income (@ 10 birr/kg)	13,400	13,100	12,800
Straw yield (t/ha)	3.5	3.9	4.5
Straw income (@ 250 birr/t)	875	975	1,125
Total income (birr)	14,275	14,075	13,925
Costs			
Seed (@ 14.40 birr/kg)	80	160	240
Urea (@ 6 birr/kg) 50 kg/ha	300	300	300
DAP (@ 7.70 birr/kg) 125 kg/ha	962.5	962.5	962.5
Total variable cost	1,342.5	1,422.5	1,502.5
Net benefit (birr)	12,932.5	12,652.5	12,422.5
Benefit to cost ratio	9.6	8.9	8.3

Table 5.6 Farmers' preferences evaluation at crop maturation, 2012

Seeding rate	Group 1	Group 2	Group 3	Group 4	Total points	Ranking
5 kg/ha + Soil	2	2	1	3	8	2nd
10 kg/ha + Soil	3	3	3	2	11	1st
15 kg/ha + Soil	1	1	2	1	5	3rd

Note: Higher points indicate a higher ranking, so 3 points means it is ranked 1st, and so on.

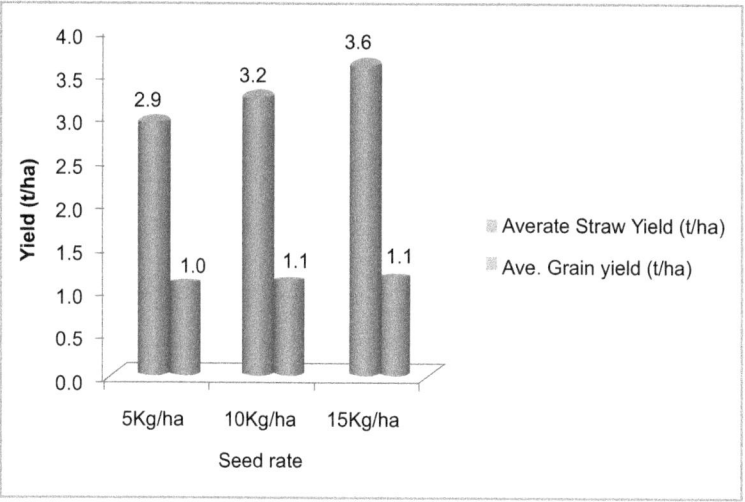

Figure 5.1 Average grain and straw yield of teff at lower seed rates using a seed spreader during 2010–12

Table 5.7 Benefit to cost ratio of farmers' preferred teff seeding rates, 2010–12

Year	5 kg/ha	10 kg/ha	15 kg/ha
2010	4.5	5.3	5.1
2011	6.0	5.9	6.3
2012	9.6	8.9	8.3
Average	6.7	6.7	6.6

Table 5.8 Farmers' crop stand preference at crop maturation, 2010–12

Year	5 kg/ha	10 kg/ha	15 kg/ha
2010	3 (1st)	2 (2nd)	1 (3rd)
2011	3 (1st)	2 (2nd)	1 (3rd)
2012	2 (2nd)	3 (1st)	1 (3rd)
Average preference	2.7 (1st)	2.3 (2nd)	1.0 (3rd)

- the decision to start the first sowing time based on the usual sowing time of the area;
- the selection of the most suitable trial field based on crop history and other locally known factors;
- preparing locally available materials for the research (soil, sieves, plastic sheets);
- active involvement and cost sharing (land preparation, preparation of pegs, guarding, seed/soil mixture preparation, knowledge transfer to non-participating farmers);
- Experiment modification: we originally planned to test teff seeding rates up to 30 kg/ha (the recommended rate). However, based on discussions with participating farmers before starting the experiment in the first year, the research stakeholders (researchers, FRG members and DAs) agreed to include a 35 kg/ha seeding rate as per local practice;
- providing ideas on the evaluation of their preferred (lower) seeding rates on larger plots;
- field evaluation at different stages, in the absence of researchers and DAs.

The FRG approach has had positive effects on the extension process in terms of the benefits of involving farmers at different stages of the trials, as their preferences fit with the scientific data analysis. This should be considered an important tool to aid the applicability of research outputs in a wider context.

Summary and conclusion

We carried out on-farm participatory evaluation of teff seeding rates using a seed spreader in Wolaita Zone, South Ethiopia, during the rainy seasons of 2010 to 2012. Combining the field results, farmers' observations at different stages of crop growth and economic analysis, 10 kg/ha mixed with soil was found to be best and most acceptable to the farmers. This rate can save 20–25 kg of seeds over the previously normal practice of 30–35 kg/ha, enough to cover the hectare and the crop is less prone to extreme weather shocks. It also promotes efficient utilization of improved and quality seed, protects farmers from extra seed costs, and is highly suitable for addressing the needs of resource-poor farmers. Nonetheless, it is also possible to reduce the seed rate further, to 5–10 kg/ha, using a spreader.

Weed infestation, which is a common problem in other major teff-growing parts of the country, was lower as a result of using lower seeding rates, and was not a major concern for FRG members in this study area. Very small landholdings in the study area could potentially allow continuous cultivation (i.e. two seasons of cultivation per year on the same land) and multiple cropping (intercrops). This frequent cultivation practice uses frequent tillage, leading to lower weed infestation. We recommend that there is further investigation of the performance of lower seeding rates with other teff varieties, rates of fertilizer application and using the row sowing method.

In conclusion, farmers' participatory research is a viable and valuable tool for passing technologies to farmers. Lowering the teff seeding rate using seed spreaders for teff cultivation is an easy, cheap and adaptable technology for farmers. As indicated in Adesanwo et al. (2009), unlike the conventional approach of passing down information to farmers where farmers' participation in research has been very poor, with variable take-up and results, farmers' participation in this experiment has succeeded in bringing together farmers, site extension workers and researchers to share ideas, and plan, evaluate and refine the seed spreader technology for teff cultivation, with benefits to everyone involved.

Acknowledgements

The authors gratefully acknowledge the Japan International Cooperation Agency FRG II project for funding this study. The staff of the Plant Science Department of Wolaita Sodo University, especially Harko Halala, Getachew Kefeta, Sara G/Meskel, Ermyas Elka and Adera Sisay, are gratefully acknowledged for data collection. We also acknowledge the development agents of Edo *kebele* (Tariku and Habtamu), the trial farmers (Ato Matalo Mana and Ato Markos Yeliso and their families) for providing us with their land for the experiment, and finally the FRG members of Edo *kebele*.

We would like to record our special thanks to Gifole Gidago for his local translation services: *Wolaitigna*.

About the authors

Fanuel Laekemariam is Assistant Professor in Agronomy and PhD in soil science at the College of Agriculture, Wolaita Sodo University, Ethiopia. He is a doctoral candidate in Soil Science, College of Agriculture and Environmental Sciences (CoAES), Haramaya University, Ethiopia.

Gifole Gidago is Assistant Professor in Soil Science at the College of Agriculture, Wolaita Sodo University, Ethiopia.

Wondimeneh Taye is a lecturer in the College of Agriculture, Wolaita Sodo University, Ethiopia, and a doctoral candidate in plant pathology at the College of Agriculture and Environmental Sciences Haramaya University, Ethiopia.

Note

1. A *kebele* is an administrative division in Ethiopia, equivalent to a village.
2. Extension workers at the village level in Ethiopia are called development agents.

References

Abule, E., Tadesse, A., Kebede, T. and Shenkute, B. (2011) 'Forage Seed Production and Multiplication through Farmers' Research Group in Adami Tulu and Arsi-Negelle Districts'. In FRG II. *Improving Farmers Access to Seed. Empowering Farmers' Innovation Series No. 1* pp. 90-101. EIAR/FRG II, Ethiopia.

Adesanwo, O.O., Adetunji, M.T., Adesanwo, J.K., Osiname, O.A., Diatta, S. and Torimiro, D.O. (2009) 'Evaluation of traditional soil fertility management practices for rice cultivation in southwestern Nigeria'. *American-Eurasian Journal of Agronomy*, 2(2): 45-49.

Anwar, P., Juraimi, A.S., Puteh, A., Selamat, A., Man, A. and Hakim, A. (2011) Seeding method and rate influence on weed suppression in aerobic rice. *African Journal of Biotechnology*, 10(68): 15259-71.

CSA (Central Statistical Agency) (2013) *Agricultural Sample Survey 2012/2013 (2005 E.C.). Volume I Report on area and production of major crops (Private peasant holdings, meher season)*. Statistical Bulletin 532. Addis Ababa, Ethiopia.

Ethiopian Seed Enterprise (ESE) (2001) *Crop Varieties Bulletin*. Addis Ababa, Ethiopia.

FRG extension materials series, No. 19 (n.d.) *Trials of Tef Production Conducted by Farmers Research Group*, Melkassa Agricultural Research Center, Ethiopia.

Fufa, H., Tesfa, B., Hailu, T., Kebebew, A., Tiruneh, K., Abera, D. and Seyefu

Ketema (2001) In Hailu, T., Getachew, B. and Sorrells, M. (eds.) *Narrowing the Rift. Tef research and development. Proceedings of the International Workshop of Tef Genetics and Improvement*, Debre Zeit, Ethiopia,16-19 October. 2000. EIAR, Addis Ababa, Ethiopia, pp. 167-76.

Gobeze, L., Waga, M., and Legese, H. (2007) 'Effect of varieties and seeding rates of triticale (*Triticosecale wittmark*) in different agro-ecologies of Southern Ethiopia', 8, pp. 41-44, *African Crop Science Society. Printed in El-Minia, Egypt* Tefer and Seyfu Ketema (2001) Production and Importance of Tef in Ethiopian Agriculture. In Hailu T., Getachew B. and Sorrells M. (eds) *Narrowing the Rift. Tef research and development. Proceedings of the "International Workshop of Tef Genetics and Improvement"*, Debre Zeit, Ethiopia.16-19 Oct. 2000. EIAR, Addis Ababa, Ethiopia, pp. 3-8.

Ketema, S. (1997) *Tef. Eragrostis tef (Zucc.) Trotter. Promoting the conservation and use of underutilized and neglected crops*. Institute of plant genetics and crop plant research, Gatersleben/International Plant Genetic Resources Institute (IPGRI), Rome.

Hall, M. H. and Cherney, J. H. (2010) *Increased teff seeding rates in the northeast region of the United States increases forage yield* [Online]. Forage and grazing lands. <http://dx.doi.org/10.1094/FG-2010-0802-01-BR>.

Lacey, T. and Llewellyn, C. (2005) *Eragrostis tef as a specialized niche crop*. No. 42/2005. www.agric.wa.gov.au.

Ministry of Agriculture and Rural Development (2007) *Crop Variety Register*. Issue No. 10. Addis Ababa, Ethiopia.

Naturland, V. (2002) *Organic Farming in the Tropics and Subtropics*. Exemplary Description of 20 Crops, Sesame, Germany, www.naturland.de V. – 1st edition.

Nishikawa, Y. (2011) Diversity of frameworks for understanding agro-biodiversity. Why do seeds matter? In FRG II. *Improving farmers access to seed*, pp. 9-20, Empowering farmers' innovation series No. 1, EIAR/FRG II, Ethiopia.

Owuor, B. O., Gudu, S and Niang, A. (2001) 'Direct seeding of *Sesbania sesban* for green manure in agro forestry systems – a short communication', *Agro forestry Systems*, 52, pp. 23-25.

Rathore P.S. (2001) *Techniques and management of field crops production*. Agrobis, New Delhi.

Statistical Analysis System (1997) *Statistical Methods*, SAS Institute Inc., Cary, NC.

Southern Nation, Nationalities, and People Regional State (2007) *Regional Statistical Abstract 1998E.C (2005-06)*. Bureau of Finance and Economic Development, Division of Statistics and Population, Hawassa.

Stallknecht, G.F., Gilbertson, K.M. and Eckhoff, J.L. (1993) Tef: Food crop for humans and animals. In: Janick J. and Simon, J.E. (eds.), *New crops*, pp. 231-34. Wiley, New York.

Tefera, H. and Belay, G (2006) 'Eragrostis tef (Zuccagni) Trotter'. In: Brink, M. & Belay, G. (eds). *PROTA 1: Cereals and pulses/Céréal esetlégumessecs*. [CD-Rom]. PROTA: Wageningen.

Tefera, H., Belay, G. and Sorrells, M. (2001) *Proceedings of the International Workshop of Tef Genetics and Improvement*, Debre Zeit, Ethiopia,16-19 October 2000. EIAR, Addis Ababa, Ethiopia.

Twidwell, E.K., Boe, A. and Casper, D.P. (1991) *Tef: a new annual forage grass for South Dakota*. Ex 8071. Cooperative Extension Service. South Dakota State University, Brookings.

Yadeta, K., Ayele, G. and Negatu, W. (2001) 'Farming systems research on tef: smallholders' pro production practices'. In Hailu, T., Getachew, B. and Sorrells, M. (eds.) *Narrowing the rift. Tef research and development. Proceedings of the International Workshop of Tef Genetics and Improvement*, Debre Zeit, Ethiopia, 16-19 October. 2000, pp. 9-24. EIAR, Addis Ababa, Ethiopia.

Yenesew, Y., Shultz, B., Abraham, Yihun, M. And Tekulu, E. (2010) 'Agricultural productivity optimization in water scarce semi-arid regions of Ethiopia' in Proceedings of the 61st International Executive Council Meeting and the 6th Asian Regional Conference of International Commission on Irrigation and Drainage (ICID), 10-16 October 2010, Yogyakarta, Indonesia.

CHAPTER 6
Participatory evaluation of selected fish processing and preservation technologies: the case of Lake Tana, Ethiopia

Shewit Gebremedhin, Markos Budusa, Adamu Yimer, Minwyelet Mingist, Dereje Tewabe and Zerihun Nigussie

Abstract

This study was conducted from March 2013 to January 2014 to evaluate and compare selected technologies for quality fish produce. Two trial sites were selected based on the local experience of traditional fish drying, market accessibility and better fish catches. For each site, one fishers' research group with 15 members was selected. The moisture content of each treatment sample was determined. Four hedonic scales were used to evaluate the processed samples organoleptically. The microbial load of processed samples of three species was carried out, following standard procedures. Using a solar tent reduced the moisture content to 18.0%, 20.8% and 19.0% at Gorgora and 22.0%, 18.2% and 19.8% at Nabega for Oreochromis niloticus, Clarias gariepinus and Labeobarbus respectively. When tested organoleptically, fish products processed using a solar tent were judged the most acceptable. The mean difference of microbial load count between the tested technologies (solar tent dry, rack dry, smoking and traditional dry) was statistically significant ($p<0.05$). The fish samples processed in solar tents had a lower microbial load than others. In conclusion, the use of solar tents should be extended.

Keywords: Ethiopia, fishers' research groups, FRGs, microbial load, moisture content, organoleptic test, solar tent

Introduction

Fish is an essential food resource, providing high-quality animal protein for over one billion people worldwide (Manasi et al., 2009). In Africa, 35 million people depend on the fisheries sector for their livelihoods (Davies and Davies, 2009). Lake Tana, the largest lake in Ethiopia, creates jobs for 3,514 fisher households (ANRSLRDPA, 2011). The Amhara region earned 65 million birr (4,062,500 USD) in 2011 from Lake Tana fisheries and 14 million birr (875,000 USD) from fish products exported to Sudan (ANRSLRDPA, 2011).

The traditional fish-processing method of gutting and kench salting is the only method in use around Lake Tana. The process is locally called *majel*. The fish is dried by hanging it over ropes or sticks without any protection from flies. It takes a week to dry. The product is prepared this way for export to Sudan and is not consumed by the local community. Fish is one of the most perishable food products (Ojutiku et al., 2009). Spoilage occurs owing to the presence of enzymes, bacteria and the chemical oxidation of fat. There are considerable fish harvest losses owing to inadequate handling, processing and marketing procedures. Mohammed (2011) reported that the post-harvest loss of Lake Tana fish was 30 per cent. The high post-harvest losses and the low price of the product adversely affects fishers' livelihoods. Though there has been limited experimental research on improved fish-drying technologies, some research was carried out by Tessema et al. (2008) and Yimer (2012) at Lake Tana to attempt to select appropriate technologies to reduce post-harvest losses, but these were not evaluated by the target community.

Trying to evaluate fish-drying technology without the involvement of the fishing community is very unlikely to address the problems. Therefore, the Farmer Research Group (FRG) approach was employed for this research, with the objective of evaluating the drying rack, solar tent, smoking and traditional drying methods to select the most appropriate drying technology to reduce post-harvest losses.

Materials and methods

Description of trial sites

Fogera and Dembia districts were chosen, from which the *kebeles* (villages) of Nabega and Gorgora were selected, based on their experience of traditional fish drying and marketing to Sudan and their high fish catches. Gorgora and Nabega are 245 and 95 kilometres from Bahir Dar respectively.

Selection of fishers' research group members

From the two trial sites, a research group comprising nine women and six men was assembled. The selection was carried out in collaboration with the agricultural development agents (DAs) and *kebele* administrators. First the DAs and *kebele* administrators identified 60 and 33 fishers in Nabega and Gorgora from 379 and 175 fisher households respectively (ANRSLRDPA, 2011), who were willing to work with researchers as members of an FRG and had five years or more experience in harvesting and processing fish. The identified fishers were given some initial training by researchers, DAs and *kebele* administrators about the planned research and what commitment was required from the members of the group. This included providing samples, constructing and evaluating the technologies, keeping the technologies safe and secure, and collecting data (i.e. length of drying time and the price of the product). The members were also informed that there would be no expenses paid and that the samples would be provided free. After the training, some fishers from Nabega and Gorgora lost interest. Fifteen fishers were finally selected from each *kebele*. Those who had shown a higher level of commitment to undertake the activities and were happy to contribute to the research. They all had seven years or more experience of fishing and fish processing.

Preparation of fish

Labeobarbus, *Clarias gariepinus* and *Oreochromis niloticus* were used for processing. For solar tent and rack drying, the samples were first washed, skinned, gutted and filleted on a clean plastic sheet using knives. The fish was then soaked in 16 per cent (160 grams per litre) brine solution for 30 minutes and drained until the surface moisture evaporated. For smoking, the sample was washed, gutted and soaked in a 16 per cent brine salt solution until the eye colour had changed to white and then drained until the surface moisture evaporated.

Drying technologies

Traditional drying. The fish for traditional drying was prepared by gutting, kench salting and hanging over ropes or sticks under shade (see Figure 6.1a). The height of the rope or stick from the ground was not fixed. The amount of salt usually added probably varies, but FRG members used a range of 1–2 handfuls of solid salt for a single fish, depending on its size.

Rack drying. The rack drier was made from wood and plastic rope resembling a bed-like construction (Tessema et al., 2008; Yimer, 2012). It was 1.4 m in length, 800 mm wide and 400 mm high. The fish sample was arranged on the rack, which stood on the ground in the open air. The fish were dried by direct sunlight and by movement of air.

Solar tent drying. The solar tent driers were made from wood and transparent polythene sheeting, as shown in Figure 6.1b. They have a vermin-proof ventilation panel made from wire mesh. The floors were covered with black plastic sheeting, which absorbs heat. The fish were arranged on the rack and put inside the drying tent.

Smoking. The smoking unit was a rectangular structure, 1.8 m high and 700 mm wide and long. After the samples were washed and soaked with salt water, as with the other treatments, the samples were smoked using *Olea europaea*

Figure 6.1 The traditional drying method (a) and solar tents (b)

and *Brucea antidysenterica*. The *Labeobarbus* and *O. niloticus* were smoked for four hours and *C. gariepinus* for six hours, all being turned at intervals. After being smoked, the fish were cooled for 10–15 minutes.

Drying duration. The length of time the fish samples were dried was recorded by researchers and FRG members. The moisture content was checked using sensory criteria: texture, colour, and odour. During checking, gloves were used to prevent contamination of the fish sample. The criteria for a fish being dried vary depending on the type of fish: dried Nile tilapia (*Oreochromis niloticus*) has a light yellow colour with neutral odour; *Labeobarbus* has a light, whitish colour and neutral odour; whereas the African catfish (*Clarias gariepinus*) has a dark red colour and a fairly typical fishy odour (Huss, 1995). In addition, properly dried fish are difficult to bend, tending to break easily. The atmospheric and tent temperature was measured using a thermometer three times a day (at three-hour intervals: 10 am, 1 pm and 4 pm). The maximum temperature was noted.

Participation of the FRG members.

FRG members participated in the installation of the fish-processing technologies, in providing the fish samples, keeping the installed technologies safe and secure, carrying out the preliminary processing (skinning, gutting, washing, filleting, slicing and brining), collecting data – such as length of drying times and the price fetched at market, and evaluating the success of the different procedures and technologies using their own experienced observation. After the evaluation, FRG members agreed to collect further data, such as the impact of the solar tent (which the members chose as the best solution) on pricing and other perceptions concerning the use of the solar tent.

Data collection. Moisture content analysis. Initially 3 kg of fish from each species was dried using the selected treatments and then a sample of 25 g was randomly taken from each treatment and species and placed in an oven. The weight change of the samples was monitored until the weight became constant. The moisture content was calculated as the difference between the initial and final (dry) weight [$(W_i - W_f)/W_i$]*100 (Clucas, 1990).

Organoleptic test. One sample from each species, processed by each of the selected treatments, was tested organoleptically by the FRG members. They evaluated the quality of processed fish without first being informed which sample was processed using which treatment to offset any possible bias. The test used a four-point hedonic scale: 3 being 'like extremely', 2 being 'like slightly', 1 being 'dislike slightly' and 0 being 'dislike extremely' (modified from Clucas, 1990).

Microbial test. From the total 3 kg sample of each species, processed by each selected treatment, 25 g samples were taken aseptically. These samples were individually transferred to sterile plastic bags. The samples were placed in sterile, standard stomacher bags containing distilled water and peptone water with a dilution ratio of 1:100, then blended and homogenized for 1–2 minutes in a stomacher. Samples were serially diluted (1:10) and spread onto various media depending on the type of micro-organism to be tested (Bridson, 2006). Then, the number of colonies were counted and changed into colony forming unit of sample. The *Staphylococcus* species, *Enterobacteriaceae*, and yeasts and moulds were identified using mannitol salt agar (MSA), violet red bile agar (VRBA), plate count agar (PCA) and potato dextrose agar (PDA) respectively. The first two were incubated at 37°C and the latter two at 35°C for one to two days.

Data analysis

Descriptive statistics were used to analyse the mean and standard deviations of microbial load of the three fish species with respect to each processing method. One-way analysis of variance (ANOVA) was used to determine the statistically significant difference of means among the selected fish-processing technologies. The statistical Software Package for Social Science (SPSS) version 16.0 was used to manage and analyse the data.

Results and discussion

Drying duration

Fish dried in a solar tent took on average 1.5 days for *O. niloticus*, 2 days for *Labeobarbus* and 2.5 days for *C. gariepinus*. Rack drying took on average 2.5 days for *O. niloticus*, 3 days for *Labeobarbus* and 4 days for *C. gariepinus*. The traditional drying method took one week or longer to dry. The time taken for smoking was 4, 4.5 and 6 hours for *O. niloticus*, *Labeobarbus* and *C. gariepinus* respectively. The temperature inside the solar tent was 56°C and 50°C, whereas the atmospheric temperature was 27°C and 25°C at Nabega and Gorgora respectively. This could be accounted for by the heat-trapping capacity of the tent and black plastic layer. The ventilation unit would also reduce humidity inside the tent. These conditions caused the sample to have lower moisture content and to dry faster than other drying methods. The longer drying time for traditionally dried fish was because the fish had been only gutted. Yimer (2012) and Tesema et al. (2008) reported 1.5 and 2.5 days taken to dry *O. niloticus* and *C. gariepinus* respectively using a solar tent.

Moisture content

The average moisture content of *C. gariepinus* (21.4%), *Labeobarbus* (19.4%), and *O. niloticus* (18.1%) dried by solar tent was lower than that recorded by *using* a rack (27.8%, 23% and 22.4%), smoking (63.6%, 66% and 61.6%) and traditional drying (64.1%, 39.2% and 37.2%) respectively. Using a solar tent reduced the moisture content to below 25 per cent, where the growth of bacteria and moulds was suppressed. The low moisture content is important in reducing autolytic activity. The result of this experiment was in agreement with Tessema et al. (2008), Degebassa (2010) and Yimer (2012), in that solar drying resulted in a lower moisture content than both rack and traditional drying methods.

Organoleptic test

All the FRG members found the fish to be acceptable when testing organoleptically the samples dried in the solar tent (see Table 6.1). This is because the solar tent had protected the samples from flies, dust and other contaminants. Moreover, the higher temperature inside the tent dried the sample faster, resulting in a better texture. Similarly, Tesema et al. (2008) and Yimer (2012) reported that solar tent-dried fish were more organoleptically acceptable than rack-dried fish. The smoked fish was popular because it has a very good flavour. FRG members rejected fish dried by the traditional method because of its bad odour, dusty colour and fragile texture.

Microbial analysis

The microbial load of the solar-dried fish was below the standard of the International Commission on Microbiological Specifications for Foods (ICMSF, 1986) (see Table 6.2). The acceptable maximum limit for *Staphylococcus* species, total plate count of *Enterobacteriaceae*, and yeast and mould count is 10^7, 10^4 and 10^6 cfu/g respectively.

The total plate count (TPC) showed significant differences ($p \leq 0.05$) between the drying methods. A higher TPC was found on fish samples processed by traditional (2.13×10^8) and rack (6.48×10^6) drying methods. This could be accounted for by the lower temperature during drying, exposure to the atmosphere, and higher microbial load of the raw material. A lower TPC was scored for fish samples processed by the solar tent drying method, 2.27×10^5 (see Table 6.2). This can be accounted for by the cumulative effect of brine salting, higher temperatures and protection from environmental contamination. In line with this finding, Ahmed and Eltegani (2012) reported that the bacterial load of dried fish decreased in line with the removal of water level to below 25 per cent – the level needed for microbial growth and enzyme activity.

The initial TPC of the smoked fish samples was low because of brining and the anti-microbial effect of smoke. This result is in agreement with Yimer (2012). However, after 15 days of storage, the microbial count was higher than the level normally expected. This is explained by the fact that the smoked fish had a high moisture content. Similarly, the growth of micro-organisms in

Table 6.1 Sensory evaluation of fish samples processed by selected methods

Treatments	Site	Tests	Dislike extremely	Dislike slightly	Like slightly	Like extremely
Solar dried	Gorgora	Odour				15
		Colour				15
		Texture				15
		Taste				13[i]
	Nabega	Odour				15
		Colour				15
		Texture				15
		Taste				12[ii]
Rack dried	Gorgora	Odour		5	10	
		Colour		9	6	
		Texture		5	10	
		Taste[1]		10	2[iii]	
	Nabega	Odour		12	3	
		Colour		12	3	
		Texture		12	3	
		Taste		10	3[iv]	
Traditionally dried	Gorgora	Odour	15			
		Colour	15			
		Texture	15			
		Taste[v]	*	*	*	*
	Nabega	Odour	15			
		Colour	15			
		Texture	15			
		Taste[v]	*	*	*	*
Smoked	Gorgora	Odour			3	12
		Colour			3	12
		Texture		2	3	10
		Taste				15

Note: Evaluated by 15 FRG members at each site. Those who refused to test the taste did so because of a lack of experience in eating the dried fish product, but all FRG members refused to test the taste of traditionally dried fish owing to the poor quality of the product.

[i] Two fishers refused to test this treatment.
[ii] Three fishers refused to test this treatment.
[iii] Three fishers refused to test this treatment.
[iv] Two fishers refused to test this treatment.

Table 6.2 Microbial load of fish samples with each treatment

Media	Treatment	No. of samples	Mean ± standard error (cfu/g)
PCA	Solar	12	$2.27 \times 10^5 \pm 2.7 \times 10^4$
	Rack	12	$6.48 \times 10^6 \pm 4.2 \times 10^5$
	Smoked	6	$1.49 \times 10^5 \pm 8.6 \times 10^3$
	Traditional	12	$2.13 \times 10^8 \pm 8.2 \times 10^7$
PDA	Solar	12	$2.5 \times 10^5 \pm 2.4 \times 10^4$
	Rack	12	$5.7 \times 10^6 \pm 2.9 \times 10^5$
	Smoked	6	$1.48 \times 10^5 \pm 7.8 \times 10^3$
	Traditional	12	$8.8 \times 10^7 \pm 1.8 \times 10^6$
MSA	Solar	12	$3.18 \times 10^5 \pm 1.0 \times 10^5$
	Rack	12	$6.06 \times 10^6 \pm 2.9 \times 10^5$
	Smoked	6	$2.11 \times 10^5 \pm 6.5 \times 10^3$
	Traditional	12	$1.36 \times 10^8 \pm 6.2 \times 10^7$
VRBA	Solar	12	$3.71 \times 10^3 \pm 1.9 \times 10^3$
	Rack	12	$1.06 \times 10^6 \pm 2.1 \times 10^5$
	Smoked	6	$2.09 \times 10^3 \pm 5.6 \times 10^2$
	Traditional	12	$3.79 \times 10^6 \pm 1.8 \times 10^5$

MSA: mannitol salt agar, VRBA: violet red bile agar, PCA: plate count agar, PDA: potato dextrose agar

Note: Where samples for smoking number 6, it is because the test was carried out in one site only.

the smoked fish depended on the amount of water which had been expelled (Oyewole et al., 2006). Moreover, smoking is only effective in controlling surface spoilage (Hilderbrand, 1992).

Shelf life

Fish samples dried using a solar tent were in the acceptable range until the last evaluation taken at 75 days of storage (see Table 6.3). However, the fish samples dried using the traditional method, smoking and drying rack were rejected at zero, 15 and 30 days of storage respectively. The drier the fish, the longer will be the shelf life.

Socio-economic analysis of the selected technologies

The effect of technology on price. In Lake Tana, the price of fresh fish is set according to the species, but with regard to dried fish product, fishers set the price regardless of species. The price of mixed product of *Labeobarbus* and

Table 6.3 Effect on shelf life of microbial load of solar-dried fish

Media	Storage time	C. gariepinus	Labeobarbus	O. niloticus
PCA	0 days	$1.69 \times 10^5 \pm 1.91 \times 10^4$	$2.67 \times 10^5 \pm 4.31 \times 10^4$	$2.46 \times 10^5 \pm 6.36 \times 10^4$
	15 days	$1.96 \times 10^5 \pm 1.36 \times 10^4$	$3.08 \times 10^5 \pm 3.21 \times 10^4$	$2.92 \times 10^5 \pm 6.44 \times 10^4$
	30 days	$2.80 \times 10^5 \pm 2.95 \times 10^4$	$3.71 \times 10^5 \pm 2.30 \times 10^4$	$3.75 \times 10^5 \pm 3.40 \times 10^4$
	45 days	$3.74 \times 10^5 \pm 6.40 \times 10^4$	$4.26 \times 10^5 \pm 5.02 \times 10^4$	$3.98 \times 10^5 \pm 3.97 \times 10^4$
	60 days	$3.25 \times 10^5 \pm 1.15 \times 10^4$	$3.46 \times 10^5 \pm 1.33 \times 10^4$	$2.93 \times 10^5 \pm 2.96 \times 10^4$
	75 days	$3.72 \times 10^5 \pm 1.64 \times 10^4$	$3.58 \times 10^5 \pm 6.81 \times 10^4$	$3.63 \times 10^5 \pm 5.07 \times 10^4$
PDA	0 days	$2.65 \times 10^5 \pm 4.98 \times 10^4$	$2.06 \times 10^5 \pm 3.65 \times 10^4$	$2.91 \times 10^5 \pm 3.69 \times 10^4$
	15 days	$3.11 \times 10^5 \pm 5.63 \times 10^4$	$2.30 \times 10^5 \pm 3.38 \times 10^4$	$3.20 \times 10^5 \pm 4.96 \times 10^4$
	30 days	$3.31 \times 10^5 \pm 2.75 \times 10^4$	$3.41 \times 10^5 \pm 2.76 \times 10^4$	$3.62 \times 10^5 \pm 4.49 \times 10^4$
	45 days	$3.55 \times 10^5 \pm 4.55 \times 10^4$	$3.46 \times 10^5 \pm 5.63 \times 10^4$	$4.05 \times 10^5 \pm 7.97 \times 10^4$
	60 days	$2.93 \times 10^5 \pm 5.93 \times 10^4$	$3.28 \times 10^5 \pm 5.31 \times 10^4$	$3.28 \times 10^5 \pm 2.08 \times 10^4$
	75 days	$3.30 \times 10^5 \pm 6.24 \times 10^3$	$4.05 \times 10^5 \pm 2.85 \times 10^4$	$3.63 \times 10^5 \pm 2.26 \times 10^4$
MSA	0 days	$2.33 \times 10^5 \pm 9.36 10^4$	$5.36 \times 10^5 \pm 2.73 \times 10^5$	$1.86 \times 10^5 \pm 6.87 \times 10^4$
	15 days	$3.75 \times 10^5 \pm 8.13 \times 10^4$	$5.84 \times 10^5 \pm 2.75 \times 10^5$	$3.53 \times 10^5 \pm 1.15 \times 10^5$
	30 days	$6.67 \times 10^5 \pm 2.96 \times 10^5$	$2.74 \times 10^5 \pm 5.97 \times 10^4$	$3.26 \times 10^5 \pm 1.00 \times 10^5$
	45 days	$7.74 \times 10^5 \pm 4.33 \times 10^5$	$3.20 \times 10^5 \pm 7.56 \times 10^4$	$6.67 \times 10^5 \pm 2.89 \times 10^5$
	60 days	$2.84 \times 10^5 \pm 5.17 \times 10^4$	$1.76 \times 10^5 \pm 2.54 \times 10^4$	$1.94 \times 10^5 \pm 6.72 \times 10^4$
	75 days	$3.25 \times 10^5 \pm 2.62 \times 10^4$	$3.21 \times 10^5 \pm 9.40 \times 10^4$	$2.35 \times 10^5 \pm 1.28 \times 10^5$
VRBA	0 days	$2.09 \times 10^3 \pm 9.18 \times 10^2$	$6.99 \times 10^3 \pm 5.92 \times 10^3$	$2.05 \times 10^3 \pm 9.70 \times 10^2$
	15 days	$2.42 \times 10^3 \pm 1.01 \times 10^3$	$7.98 \times 10^3 \pm 5.90 \times 10^3$	$1.73 \times 10^3 \pm 1.32 \times 10^3$
	30 days	$3.34 \times 10^3 \pm 1.08 \times 10^3$	$9.97 \times 10^3 \pm 6.40 \times 10^3$	$8.64 \times 10^3 \pm 3.71 \times 10^3$
	45 days	$3.12 \times 10^3 \pm 1.58 \times 10^3$	$3.03 \times 10^3 \pm 1.24 \times 10^3$	$7.02 \times 10^3 \pm 1.94 \times 10^3$
	60 days	$3.41 \times 10^3 \pm 2.31 \times 10^3$	$3.25 \times 10^3 \pm 8.44 \times 10^2$	$4.27 \times 10^3 \pm 9.64 \times 10^2$
	75 days	$5.00 \times 10^3 \pm 1.55 \times 10^3$	$4.30 \times 10^3 \pm 6.71 \times 10^2$	$2.77 \times 10^4 \pm 2.41 \times 10^4$

C. gariepinus produced by traditional drying was 8 birr/kg at both study sites. After the introduction of the solar tent, FRG members started to use the solar tent to dry fish following the local procedures of drying fish (keeping the fish heads on when they are dried) for Sudanese customers (Sudanese consumers want to check whether the product really is fish or not by seeing the heads). Amare and Taye, fishermen who were members of the FRG and owners of the treatments, reported a price of 15 and 10 birr/kg from the Nabega and Gorgora sites respectively.

Perceptions of the technology. FRG members showed strong agreement on the relevance of the solar tent (the other technologies having been rejected) in improving traditional processing practices. This was accounted for by the following attributes: 1) simplicity – the technology was perceived as simple to understand and easy to use; 2) compatibility – the technology was perceived as suitable to the resources, needs and practices of the fishers; 3) observability – the produce is attractive not only to FRG members but also to non-target groups; 4) divisibility – suitability also for using with modified versions of the drying process. During our testing period, the fish were gutted, washed, skinned, filleted, sliced and then dried using the solar tent. However, fishers didn't adopt this product since it did not have a local market and Sudanese consumers want to see the heads still in place. The fishers therefore adopted a partial process where the fish were dried using the technology but they then followed local procedures (i.e. the fish were gutted and washed only, and then dried); and finally 5) the possibility of constructing the dryers from local materials. In this case, the materials that the researchers used were expensive and not readily available in the communities where the trial took place. The FRG members were asked whether they would be able to construct the technology from locally available materials and all agreed that they could.

Perceptions of the final product. The opinions held by FRG members about the fish product resulting from using the solar tent agreed with the researchers' original hypothesis. The members didn't even want to smell the fish processed using the traditional method. They mentioned different product-related attributes as the reasons for their new preference, such as attractive colour, long shelf life, improved texture and absence of contamination from flies and dust. The Fogera plain in which the Nabega trial site was located is flooded from July to November. The fishers transport their fish either by bicycle, horse cart or on foot. Thus, during these months there is no way to take their fish to the market in Woreta town. A few people who have a motorized boat can take their product to Bahir Dar town market. The only option for the majority of the fishers is to dry their fish product and sell it after the floods have receded and they are again able to travel to the market. FRG members of the Nabega site all agreed that using the solar tent would be a solution to the seasonal problem of market access.

Contribution of FRG members to research

FRG members contributed a number of important suggestions and comments during the research period: 1) cats or dogs may damage the tent while trying to take the fish, so guarding the solar tent would be important; 2) rather than using wire mesh for the rack which rusts easily, it would be better to use plastic rope or wire mesh covered by plastic; and 3) the solar tent should be placed on level ground for stability and to prevent insects or dust entering. FRG members also raised questions such as whether the weight loss of fish processed using the tent would lower the price of the product and result in losing local consumers.

Because of the questions raised by FRG members during the first season of research activity, the weight loss and profitability of each processed product were tested during the second season. During the first season, FRG members also showed an interest in using the solar tent with the traditional local procedure of only gutting and washing the fish first, so this treatment was also tested in the second season. There were also 'field days' during the second season in which experts from zone and district fisheries, FRG members and researchers met and discussed the problem of post-harvest losses in the wider area and possible improvements which might impact on the fishers' livelihoods. On these field days the district and zone fisheries experts were frequently asked by FRG members to promote the use of the solar tent technology and help to create a local market for the fish products produced using it.

Conclusion

Participatory research has enabled researchers and FRG members to share their knowledge with each other. FRG members' evaluation and laboratory analysis both showed that the solar tent drier produces better quality fish products than the other processing methods tested. Traditionally processed fish was found to be relatively poor in quality and was rejected by FRG members' evaluation and borne out by laboratory analysis. FRG members felt there was great potential for the wider adoption of the solar tent technology, particularly if new markets could be established.

About the authors

Shewit Gebremedhin is Assistant Professor, Department of Fisheries, Wetlands and Wildlife Management, at Bahir Dar University.

Markos Budusa is Lecturer, Department of Environment, Gender and Development Studies, at Hawassa University.

Adamu Yimer is Lecturer, Department of Fisheries, Wetlands and Wildlife Management, at Bahir Dar University.

Minwyelet Mingist is Associate Professor, Department of Fisheries, Wetlands and Wildlife management, at Bahir Dar University. He finished Doctoral degree in Fisheries Science From Hokkaido University, Sapporo, Japan.

Dereje Tewabe is Director of Fisheries and other aquatic life research at Amhara Region Agricultural Research Institute.

Zerihun Nigussie is Assistant Professor, Department of Agricultural Economics, at Bahir Dar University. He finished Master of Science in Agricultural Economics from Haramaya University, Ethiopia.

References

Amhara National Regional State Livestock Resources Development and Promotion Agency (ANRSLRDPA) (2011) *Lake Tana fisheries management plan and processing manual*. Bahir Dar, Ethiopia.

Bridson, E.Y. (2006) *The OXOID manual*. 6th edition. OXOID Limited, Basingstoke.

Clucas, I.J. (1990) *Fish Handling, Preservation and Processing in the Tropics*: Part 2 (NRI), London, Tropical Development and Research Institute.

Davies, O.A. and Davies, R.M. (2009) 'Traditional and improved fish processing technologies, in Bayelsa State, Nigeria', *European Journal of Scientific Research*, 26(4), pp. 539-548.

Degebassa, A. (2010) *A comparative study on the effect of three drying methods for better preservation of fish: Management of shallow water bodies*, EFASA, Addis Ababa, Ethiopia.

Hilderbrand, K.S. (1992) *Fish smoking procedures for forced convection smokehouse*. Special Report 887, pp. 1–41, Oregon State University Extension Service.

Huss, H.H. (1995) *Quality and change in fresh fish*, FAO fisheries technical paper 348, Fisheries and Aquaculture Department, Rome.

International Commission on Microbiological Specifications for Foods (ICMSF) (1986) *Quality Assurance of Seafood*. Food and Agriculture Organization of the United Nations (FAO), Rome.

Manasi, S., Latha, N., and Raju, K.V. (2009), Fisheries and Livelihoods in Tungabhadra Basin: Current Status and Future Possibilities, Working Paper 217, Institute for Social and Economic Change, India.

Mohammed B. (2011). Assessment of motorized commercial gill net fishery of the three commercially important fish species in Lake Tana, Ethiopia, MSc thesis, Bahir Da University, Ethiopia.

Ojutiku, R.O., Kolo, R.J. and Mohammed, M.L. (2009) Comparative study of sun drying and solar tent drying of *Hyperopisus bebe* occidentals, Pakistan, *Journal of Nutrition*, 8: 955-957.

Oyewole, B.A., Agun, B.J. and Omotayo, K.F. (2006) Effects of different sources of heat on the quality of smoked fish, Helsinki, Finland, *Journal of food, agriculture and environment*, 4(2): 95-97.

Tessema, A., Demissie, S., Goshu, G., Bekele, B., Fentahun A. and Bezabih, B. (2008) Evaluation of solar tent and drying rack methods for the production of quality dried fish in Lake Tana area. Proceedings of the 3rd annual regional conference on completed livestock research activities (CLRA). Bahir Dar, Ethiopia, pp 15-26.

Yimer, A. (2012). Assessment of selected post-harvest processing techniques for quality fish product at Bahir Dar town. MSc thesis in fisheries, wetlands and wildlife management, college of agriculture and environmental sciences, Bahir Dar University, Bahir Dar, Ethiopia.

CHAPTER 7

Participatory evaluation of farmer-saved and purified seed for improved agronomic performance: wheat, South eastern Tigray

Alem G/tsadik, Kelali Haftu, Yoshiaki Nishikawa, Ibrahim Fitiwy and Taku Seo

Abstract

A participatory evaluation considering the performance of three farmer-saved wheat seed varieties and the effects of purifying them by using a salt solution was carried out in Didba Kebele of Enderta Woreda (district) over three years, from 2011 to 2013. The objective was to assess the agronomic and economic performance of the purified seed against a control and certified seed. The results of the experiment have shown that the planting of salt solution-purified saved seed has both agronomic and economic advantages over non-purified seed. The average grain yield difference of 555–673 kg/ha and an economic advantage of 4,649–5,558 birr/ha was obtained from the purified seed when compared with the non-purified seed. A quality analysis carried out on third-generation Hawi variety seed, preserved using a salt solution, against certified seed of the same variety and a control showed that the quality of farmer-saved seed can be maintained by following these simple and cost-effective methods.

Keywords: Ethiopia, Farmer Research Group, farmer-saved seed, salt solution, participatory research, purification, wheat, grain yield

Introduction

Wheat (*Triticum aestivum* L.) is the most important cereal crop worldwide and is also important in Ethiopia. In Tigray, wheat is the fourth most important crop after sorghum, finger millet, and maize. The total production of wheat during the 2011/12 cropping season was more than 29 million metric tonnes in Ethiopia, of which 200,170 metric tonnes came from the Tigray region (CSA, 2012).

The quality of the seed used is a key input for improving crop production and productivity. Increasing the quality of seed can increase the yield potential of the crop by significant quantities (FAO, 2006). High-quality seed is a major factor in obtaining a good crop stand and rapid plant development even under adverse conditions, although other factors such as rainfall, agronomic practices, soil fertility, and pest control are also crucial. Seed quality results from the genetic, physical, physiological and phytosanitary characteristics of the seeds. When seed has good phytosanitary qualities, farmers have far better prospects of producing a good crop.

In many developing countries, the informal sector – including farm-saved seed and farmer-to-farmer sales/exchange – accounts for most planted seed (Ndiema et al., 2005). Ethiopian agriculture requires over 700,000 metric tonnes of seed each year to grow cereals and pulses (IFPRI, 2010). Of this, less than 10 per cent is met by formal seed-producing enterprises (ESE, 2007). The remainder of the seed requirement is met from the informal seed system.

Farm-saved seeds have the following benefits:

- the seed required is available at the right time;
- it saves the cost of buying seed;
- it produces seed that is adapted to the local climate, soil, etc.;
- it increases income from local resources;
- it helps to improve local varieties and conserve biodiversity.

When farm-saved seed used for planting purposes is not 'clean', however, it may potentially carry several diseases, which could remain dormant in the seed between sowing and harvesting. Farm-saved seed may be infected or of low quality and may cause early infections in the crop which lead to poor germination and/or low yields. Farm-saved seed might also have other impurities, such as inert materials like soil, stones and chaff, weed seeds, immature seeds, and broken seeds. These all reduce the quality of the seed for planting.

Resource-poor farmers are not able to use certified seeds from improved varieties every year. Neither is there any formal seed-cleaning technology, such as seed-cleaning machines that could be used by farmers or other seed-producing companies in the area. Some farmers use simple homemade sieves to clean farm-saved seeds. This method doesn't, however, separate poor-quality seeds that happen to be a similar size to the normal seeds. A group of farmers who were interested in looking for improved methods of purifying farm-saved seed, together with a team of researchers and extension workers, were thus organized by Mekelle University to work in collaboration with the Woreda Office of Agriculture and Rural Development. A Farmer Research Group based research programme was initiated in 2011 to set about evaluating farmer-saved and purified wheat seed for its agronomic performance.

Objectives

The objectives of this farmer-led experiment were to:

- evaluate the agronomic performance of farmer-saved and purified wheat seeds in farmers' fields;
- estimate the cost–benefit ratio of farmer-saved and purified seeds;
- introduce appropriate seed-cleaning technologies to FRG members.

Materials and methods

The trial sites

The study was conducted in Didba *kebele*[1] of Enderta *woreda*[2] in Tigray. Tigray, the northernmost regional state of Ethiopia, lies between 12° N and 15° N latitudes and 37°10' E and 40°10' E longitudes. The study site is located about 25 kilometres south of Mekelle, the regional capital. It has an altitude of 2,245 metres above sea level and lies at 13°36' N latitude and 39°54' E longitude. The *woreda* has a population of 114,297 and an area of 93,048 hectares. The area experiences a uni-modal type of rainfall (one rainy season in a year), which falls in the months from July to September. Cereal crops-based mixed farming is the mainstay of the farming community in the area. Wheat and barley are the first and second most important cereal crops in Enderta *woreda* covering 9,500 and 9,600 ha respectively. A survey carried out in Didba *kebele* in 2011 showed that the productivity of wheat is as low as 1,600 kg/ha. Similar results were also reported from Enderta *woreda*. The main factors contributing to such low productivity include low rainfall coupled with poor distribution, poor crop management practices and low input utilization, among others. Of all the factors affecting crop yields, the use of quality seed is paramount.

Experimental design and treatments

This participatory research was conducted from 2011 to 2013 for three years. Two treatments, namely seed purified with a salt solution, and non-purified, farmer-saved seed, were tested with three wheat varieties over the three-year period, except during the first year when only two seed varieties were used. In the 2011 cropping season, evaluation of the salt solution-purified seed of two wheat varieties, Kubsa (HAR 1685) and Hawi (HAR 2501), was conducted in nine trial farmers' fields. A third local wheat variety, Shehan, was added during the second and third years as participant farmers were interested to observe the effect of purification on the performance of the local variety. Shehan has been used for decades in the area. It is known by its white-coloured grains and the good-quality breads baked from it, and fetches a premium price in the market. The variety Hawi was released by Kulumsa Agricultural Research Centre in 1999. This variety is drought-tolerant and gives high yields. Kubsa variety was released by Kulumsa Agricultural Research Centre in 1995. Kubsa is praised for its high productivity, white grains and good baking quality. Thus, the test varieties were selected for their higher yield and preference shown for them by farmers in the area. The three varieties were planted in 11 trial farmers' fields, with each of the varieties planted in three or four fields. Seeds harvested from plots planted with purified and non-purified seeds during the first season were stored separately. The same procedure was used to further purify the first year's pure seeds for the second year and the second year's pure seeds for the third season. The seed was weighed, labelled and stored in dry and ventilated areas in the participants' houses until the next cropping season.

Biophysical data, including biomass yield, were collected based on farmers' suggestion of the importance of straw for livestock feed in the area. Crop samples for grain and biomass yield determination were taken from one 2 m by 2 m quadrat from each of the FRG trial farmers' fields. Other yield components, such as plant height and spike length, were measured from 10 randomly selected plants per plot and their mean averages were considered for comparison across the treatments. Three grain samples were taken from non-FRG farmers who planted certified seeds of the same wheat variety as the FRG farmers for purity analysis. Socio-economic data, such as the cost of inputs used, labour utilized and the perceptions of farmers on the introduced technology, were also collected. As with the agronomic data, the socio-economic data were collected from the trial FRG farmers involved in the research. The data were analysed by researchers and, from the outset, the research outputs were shared with participant farmers. Not only the results, but also the whole process of the FRG experiments was evaluated by the farmers, researchers and extension workers. Joint research plans for the forthcoming seasons were made following the evaluation of the experimental outputs of the previous cropping season.

Seed purification

All the seed for the three wheat varieties used in the experiment was supplied by the participating farmers. The seed had been saved and reused for four to six years by the farmers. As these farmer-saved seeds had been physically contaminated at various stages of crop production, cleaning the seed using an appropriate seed-cleaning method was crucial. Among the various methods that can be used to purify seeds, the salt solution purification method was chosen for use in this experiment. This method was suggested after looking at experience from Japan in the purification of farmer-saved rice seeds. In this method, 2.4 kg of table salt was added to a 10-litre bucket, which was then filled with water. The salt and water was stirred well until the salt was completely dissolved. Half of the wheat seed from each variety was added separately to the salt solution for purification. The seed was stirred until all spoiled seeds and other impurities floated on the surface of the solution (see Figures 7.1–7.3). Once the impurities were removed, the salt-water solution was poured into another bucket, leaving behind the purified seed. The wheat seed was then washed three times with clean water to remove all the salt. The seeds were spread on a plastic sheet and left to dry in the shade, with regular agitation to facilitate even drying. After about 15 minutes, the seed was dry and ready to be planted.

Seed purification and planting methods

A uniform seeding rate of 150 kg/ha was used for all three wheat varieties and treatments used in the trial. Farmers used row planting for both purified and non-purified treatments. Plot sizes ranging from 100 square metres to 1,250 m^2 were used for each treatment during the experimental period.

Figure 7.1 Farmer pouring seeds into salted water

Figure 7.2 Spoiled seeds and other foreign material floating on the surface of the salt solution

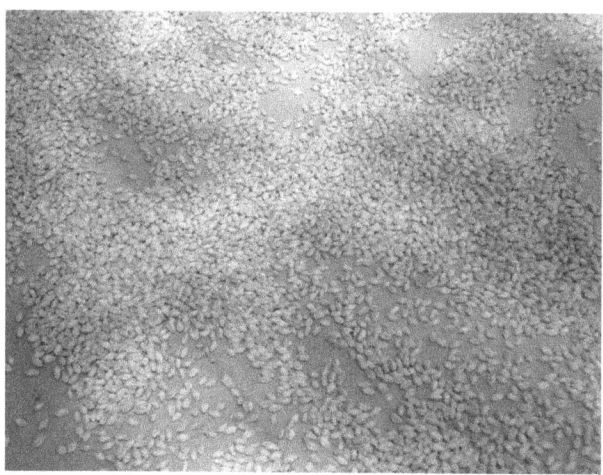

Figure 7.3 Purified seeds being dried in the shade before planting

After observing the positive results from the first season, the farmers and researchers discussed how to optimize the benefits of the research, leading to the utilization of the salted water remaining after the seeds were purified for livestock feed. Initially the salted water was fed directly to the livestock, but later the farmers decided to mix the solution with straw first, then feed it to their livestock.

Seed quality analysis

A seed quality analysis was carried out to evaluate the quality of the salt solution-purified seeds after three generations, against the control (non-purified seeds from the third generation) and certified seeds of the same variety – Hawi. Hawi was selected for the analysis since certified seed of the second improved variety, Kubsa, was not available in the market. To analyse the purity of the three seed types, samples were taken from each seed lot. The working samples from the composite samples were then separated into pure seeds, other seeds and inert matter. For the germination test, 100 kernels were taken from the pure seed of each of the three seed types. Three tests were made for each seed type. After the end of the seven-day germination period, normal seedlings were separately counted from the non-germinated seeds. Similarly, 1,000 kernels were taken at random from the seed lots of each of the three types of seed and weighed to determine the 1,000 grain weight in grams.

Selection of FRG members and team organization

When Mekelle University (through the College of Dryland Agriculture and Natural Resources) agreed to work on collaborative research with Japan International Cooperation Agency and the - FRG II project, a team from the university met staff of the Enderta *Woreda* Office of Agriculture and Rural Development and raised the idea of working with farmers following FRG approaches on maintaining the quality of farmer-saved seeds. The idea was welcomed by the office and they suggested working with farmers from Didba *kebele*. This took account of its potential for wheat production, accessibility of the *kebele* and the willingness of the farming community to learn from others. The research team then met leaders and extension agents from the *kebele* for discussions. After being introduced to the new approach and the objectives of the trial, the Office of Agriculture and Rural Development agreed to discuss it with the community. Participant farmers were selected by the village leaders and development agents (DAs) based on their willingness and accessibility, and from both male and female farmer groups. The total number of farmers involved was 111, 159 and 170 for the 2011, 2012 and 2013 cropping seasons respectively. A joint work plan was prepared by farmers, DAs and researchers to evaluate whether or not the use of salt solution purification could improve the agronomic and economic performance of farmer-saved wheat seeds. Since the research was planned to be carried out in farmers' fields with the full participation of the farmers, trial farmers were selected also taking into account their willingness to share costs, to freely contribute farm plots for the research, and for their commitment to share their experiences with others. The number of member farmers working in each of the nine to 12 groups established during the three years ranged from 12 to 15. The members of each FRG collaborated in the planning, implementation and monitoring of the trials. The farmers developed rules for the group which included conducting meetings with other farmers every two weeks during the first year and monthly during the following two, to enable them to monitor the status of their experiment and discuss the challenges encountered. A team of researchers from Mekelle University and DAs from Didba *kebele* worked closely with the farmers. Unlike with conventional research, where researchers conduct their trials either on-station or at farmers' fields separately, the role of the researchers in this FRG research was to facilitate the process and provide training on FRG approaches, the importance of using clean seed, methods of seed cleaning and data collection techniques.

A multidisciplinary team of researchers with backgrounds in agronomy, socio-economics, crop protection and animal nutrition was organized to formulate and carry out the FRG research project. The research team, in collaboration with participant farmers and DAs, were responsible for the planning, implementation and evaluation of the study. To make the coordination of planned activities and resources more straightforward, a team leader was selected from the research team. The participant farmers were also organized into groups with a chairman and a secretary. They organized joint meetings, evaluated field activities, arranged farmer-to-farmer visits and shared experiences with others. The role of the extension workers was to help the farmers implement the trials themselves. Based on the identified gaps by farmers and extension workers in data collection techniques and row planting skills, training was organized by the research team to improve the capacity of farmers and DAs.

Collection of biophysical and socio-economic data

Biophysical and socio-economic data required for analysing the agronomic and economic performance of purified seed and non-purified seed were jointly collected by researchers and farmers. Discussions were held with participating farmers on the data to be collected and on how to collect it. Data collection sheets were developed by researchers and distributed to all the trial farmers. Practical training was given to farmers on how to measure the agreed data and the necessary measuring tools were provided. Accordingly, data on days to emergence, days to heading, days to physiological maturity, plant height, and spike length were all recorded by the farmers. The remaining parameters, such as yield (grain and biomass) and 1,000 kernel weight, were jointly measured by the farmers and researchers. Purity analysis and germination tests were made by researchers. Costs incurred in the two treatments were also collected for economic analysis. The perceptions of farmers on the treatments and the three varieties used were also collected at different stages of crop growth. The collected data was analysed using simple statistical methods.

Results

The results from 2013 showed that higher mean grain yield was obtained from the purified seed for all three wheat varieties (see Table 7.1). A 657 kg/ha higher grain yield was obtained from the plots planted with the purified Shehan variety rather than non-purified seed. Similarly, a 606 kg and 609 kg/ha yield advantage was gained from purified Kubsa and Hawi varieties respectively over the non-purified seed. The total mean grain yield obtained from plots planted with purified seed of the three wheat varieties was 622 kg/ha (14.8%) higher than that from the non-purified seed. A 4,807 kg/ha total mean grain yield obtained from the plots planted with purified seed of the three varieties in the 2013 cropping season was 2,778 kg/ha higher than the national wheat average for the 2011/12 cropping season, which was only 2,029 kg/ha. Similar results were observed during the 2011 and 2012 cropping seasons. Accordingly, higher grain and biomass yields were obtained from purified seed of the tested varieties. Although not statistically significant, the average grain yield obtained from the purified seeds was 673 kg/ha (20.3%) and 555 kg/ha (14.6%) higher than the non-purified seed for the 2011 and 2012 cropping seasons respectively. Higher plant height, spike length and number of seeds/spike were also registered from plots planted with purified seed of the tested varieties during the 2011 cropping season.

Looking at each of the three wheat varieties separately, higher grain, straw and biomass yields were obtained from plots planted with purified seed than with non-purified seed. Taking Kubsa as an example, the grain, straw and biomass yields obtained during the 2013 cropping season were 5,170, 5,784 and 10,945 kg/ha respectively against 4,563, 5,135 and 9,699 kg/ha obtained from the plots planted with the non-purified seed of the same variety. Similar results were obtained during the 2011 and 2012 cropping seasons (see Table 7.1).

The data for the 2013 cropping season have shown that average grain yields of 5,592 and 4,866 kg/ha were obtained from Hawi and Kubsa wheat varieties, significantly higher than Shehan which recorded a mean grain yield of 2,539 kg/ha. Similarly, a statistically higher mean biomass yield of 11,255 and 10,326 kg/ha was recorded from plots planted with Hawi and Kubsa varieties respectively than those planted with Shehan variety, which registered a mean biomass yield of 8,976 kg/ha (see Table 7.2).

Using analysis of variance, the interaction between seed purification and varieties for the 2013 cropping season showed statistically significant differences between plots planted to purified and non-purified seeds. A significantly higher mean grain yield of 5,896 kg/ha was obtained from Hawi variety purified seed, which was statistically similar to plots planted with

Table 7.1 Yields of purified and non-purified seeds of Kubsa, Hawi and Shehan varieties from 2011 to 2013

Year	Seed type	Kubsa			Yield Hawi			Shehan		
		Biomass (kg/ha)	Grain (kg/ha)	Straw (kg/ha)	Biomass (kg/ha)	Grain (kg/ha)	Straw (kg/ha)	Biomass (kg/ha)	Grain (kg/ha)	Straw (kg/ha)
2011	Purified	8,373	3,550	6,366	8,372	2,838	5,535	The variety was not tested		
	Non-purified	6,923	2,770	4,811	6,921	2,370	4,557	The variety was not tested		
2012	Purified	5,162	8,134	13,296	14,222	5,381	8,841	8,778		6,623
	Non-purified	4,555	7,741	12,296	11,583	4,737	6,846	7,322	1,769	5,553
2013	Purified	10,954	5,170	5,784	11,639	5,897	5,742	9,844	2,868	6,976
	Non-purified	9,699	4,564	5,135	10,872	5,288	5,585	8,109	2,211	5,898

Table 7.2 Grain and biomass yields of Hawi, Kubsa and Shehan varieties, 2013

Variety	Mean grain yield (kg/ha)	Mean biomass yield (kg/ha)
Hawi	5,592a	11,255a
Kubsa	4,866a	10,326a
Shehan	2,539b	8,976b

Note: Values connected by the same letter across a column are not statistically different, with ≤5% being statistically significant

non-purified seed of the same variety and purified seed of Kubsa variety, which recorded a mean grain yield of 5,287 and 5,171 kg/ha respectively. Plots planted with purified and non-purified seed of Shehan variety, however, showed a statistically similar but lower mean grain yield of 2,868 and 2,210 kg/ha respectively (see Table 7.3).

Unlike grain yield, significant differences were observed between purified and non-purified farmer-saved seeds for biomass yield. A statistically higher mean biomass yield of 10,812 kg/ha was obtained from the plots planted with the purified seed than from those planted with non-purified seed, which registered a mean biomass yield of 8,580 kg/ha. A mean biomass yield of 1,734 kg/ha was obtained from purified Shehan seed, higher than non-purified, and higher biomass yields of 1,255 and 7,66 kg/ha were obtained from plots planted with purified seed of Kubsa and Hawi varieties respectively, during the 2013 cropping season.

The interaction effect has shown that a significantly higher biomass yield of 11,638 kg/ha was obtained from plots planted with the purified Hawi variety. Statistically similar biomass yields of 10,954, 10,872 and 9,844 kg/ha were obtained from the purified Kubsa variety, non-purified Hawi variety and purified Shehan variety respectively (see Table 7.3). A statistically lower biomass yield of 8,109 kg/ha was obtained from non-purified Shehan variety seed (Table 7.4).

Table 7.3 Interaction effect between seed purification and varieties on grain yield, 2013

Variety	Purified	Non-purified	Mean
Kubsa	5,170ab	4,563b	4,866.5
Hawi	5,896a	5,287ab	5,591.5
Shehan	2,868c	2,210c	2,539
Mean	4,645	4,020	4,332

Table 7.4 Interaction effect between seed purification and varieties on biomass yield, 2013

Variety	Purified	Non-purified	Mean
Kubsa	10,954ab	9,699bc	10,326.5
Hawi	11,638a	10,872ab	11,255
Shehan	9,844abc	8,109c	8,976.5
Mean	10,812	9,560	10,186

Seed quality analysis

The quality analysis (purity) and germination test carried out for the third generation purified seed and certified seed of the same wheat variety has shown an almost similar effect. Accordingly, the purity of the salt solution-purified seeds (99.8%) was higher than that of the certified seeds (99.2%) of the same variety, and both have shown a similar 1,000 kernel weight of 34.2 grams and a germination rate of 99.2% (see Table 7.5).

Table 7.5 Evaluation of certified seeds and third-generation, farmer-saved purified and non-purified seeds Hawi variety

Type of seed	Purity analysis (%)	1,000 grain weight (g)	Germination rate (%)
Non-purified farmer-saved seed	99.0	32.5	98.5
Purified farmer-saved seed	99.8	34.2	99.2
Certified seed	99.2	34.3	99.2

Socio-economic analysis

The seed purification process was carried out using readily available household items such as plastic buckets, or pots of known volume. Thus, the only cost for the farmer was the salt. The results of a partial budget analysis undertaken for the 2011, 2012 and 2013 cropping seasons showed a net benefit of 4,649, 4,979 and 5,558 birr/ha respectively from use of purified seeds over the non-purified seeds. However, it was very difficult for the participant farmers during the planning stages to believe that purification of saved seeds using salt solution would have such a positive result. As soaking of wheat seed, even in ordinary water, was not common practice in the area, it was difficult to accept that salt water purification would be so effective in preventing the deterioration in the quality of farmer-saved seed. After a series of discussions, however, nine trial farmers from among the farmers willing to participate in the FRG research project were convinced to try it for one season and learn any lessons.

Another issue raised during the project inception was the possible negative impact on the soil with repeated use of salt solution-purified seed. This fear was resolved by repeated rinsing of the purified seed with clean water, until all traces of the salt were removed.

Discussion

This three-year research project, conducted with close collaboration between researchers, farmers and extension workers, has proved that participatory research is capable of generating technologies that can solve local problems, and that it increases the capacity of participant farmers to innovate. Washing purified seed with fresh water to remove traces of salt and utilizing the salted water for livestock feed are two practical examples of farmers' innovations during the research process. Participant farmers' commitment during the three-year programme remained high. They freely contributed land for the experiment and seed of the three selected varieties for the entire research period. They also contributed their labour in managing the experimental plots. The farmers were willing to share their practical experience with the researchers during FRG approach training organized by Mekelle University in collaboration with the Project for Enhancing Development and Dissemination of Agricultural Innovations through Farmer Research Groups (FRG II). Thus, FRG-based research has increased the farmers' sense of ownership of the research, as was evidenced by the efficient follow-up of the research implementation by organizing farmer meetings every fortnight. The meetings were coordinated by a chairman and secretary elected from the participant farmers. These meetings became discussion forums where the best way to manage the experimental plots and all aspects of the experiment were open to debate. Using clean seed, field monitoring during crop growth, knowing the right time to harvest, the importance of using a clean threshing floor, and the use of clean and aerated storage structures were all among issues debated and resolved during these discussions.

The use of salt solution to purify farmer-saved seeds is sustainable because of farmers' ready access to table salt at a low price. It is a simple and effective technology in the context of these communities. The technology has little or no negative effect on the environment in terms of increasing soil alkalinity, as the purified seeds are washed well with tap water and the salted water is used as livestock feed to improve the palatability and nutrition of feedstuffs such as crop residues.

Conclusion

The FRG-based research conducted on evaluating the agronomic and economic performance of salt solution-purified wheat seed proved that the quality of the seed can be maintained or even optimized over successive seasons, as evidenced by the higher agronomic and economic performance of the purified seed over the non-purified seed across the three-year period of the research. The agronomic and economic performance of the purified seed of all three wheat varieties tested was better across the board than with non-purified seed. Although the objective was not to compare the three wheat varieties used, it was observed that the performance of the varieties was not the same and farmers have clearly understood what to use for the future. Farmers' own selection criteria of the varieties were not only related to yield: baking quality and market value were also considered important, as observed in the local variety Shehan. As there had been no previous participatory research undertaken in the *kebele*, farmers felt remote from support by the researchers. The collaborative approach to this research has thus improved the relations between researchers, extension workers and farmers and between the farmers themselves. The use of the participatory approach resolved the initial worries of farmers at the outset of the research, as the farmers themselves were part and parcel of the solution.

The quality analysis made on the third-generation seed from salt solution-purified seed of Hawi variety against certified seed of the same variety and a control has shown that the quality of farmer-saved seed can be maintained while following this simple and cost-effective method. At the same time, the technology used was found to be both economically beneficial and sustainable.

Acknowledgements

The compilation of this chapter has been partly assisted by a research grant by the Japan Society for Promotion of Science (No. 26304033).

About the authors

Alem G/tsadik works in CASCAPE project funded by the Kingdom of The Netherlands. The project is based at Mekelle University to identify, document and disseminate best practices in agricultural production. He has been working as lecturer in the Department of Dryland Crop and Horticultural Sciences in Mekelle University.

Kelali Haftu works as lecturer in the department of Dryland Crop and Horticultural Sciences in Mekelle University.

Yoshiaki Nishikawa is a professor of Agricultural Economics and Resource Management at Ryukoku University, Japan. His main research theme is seed system management in Japan and African countries as well as participatory research.

Ibrahim Fitwy is an associate professor working in the department of Dryland Crop and Horticultural Sciences in Mekelle University.

Taku Seo works with the Smallholder Horticulture Empowerment and Promotion Project for Local and Up-scaling (SHEP-PLUS), a technical cooperation project between the Ministry of Agriculture, Livestock and Fisheries of Kenya and the Japan International Cooperation Agency. He worked for FRG II between 2010 and 2015 as an expert in charge of training and appropriate technology development.

Notes

1. *Kebele* is an administrative division in Ethiopia equivalent to a village.
2. *Woreda* is an administrative division in Ethiopia equivalent to a district.

References

Central Statistical Agency (CSA) (2012) *Agricultural Sample Survey 2011/12. Report on Area Production of Crops (Private Peasant Holdings, Meher Season)*, The FDRE Statistical Bulletin, Vol. 1. Addis Ababa, Ethiopia.

Ethiopian Seed Enterprise (ESE) (2007) *Report on Certified seed requirements*, Addis Ababa, Ethiopia.

Food and Agriculture Organization of the United Nations (FAO) (2006) *Seed and Plant Genetic Resources Service*. Draft publication, Rome, Italy.

International Food Policy Research Institute (2010) *Seed System Potential in Ethiopia: constraints and opportunities for enhancing production*, Washington, USA.

Ndiema, A. C., Kinyua, M. G., Mahagayu, M. C., De Groote, H., Wanyera, R., and Kamundia, J. (2005) Socio-economic factors influencing use of farm saved maize and wheat seed by farmers in Nakuru district, Nairobi, Kenya.

CHAPTER 8

Improved dairy production and changing gender roles: experience of smallholder FRGs in Melkassa, Central Rift Valley

Bedru Beshir

Abstract

Local dairy cows are low in productivity, and keeping large numbers of them is becoming difficult in Ethiopia. A crossbreed of Boran x Jersey dairy heifers was introduced to Farmer Research Groups (FRGs) using revolving funds from 2007. To assess and document the process and results, data were obtained through questionnaires, interviews and observation. The data showed that the FRG served as a platform for learning new ideas, joint planning and sharing experience. The dairy FRG was instrumental in the repayment of the revolving fund. Improved dairy cows were kept in shelters and supplied with forage and clean water, unlike the habitual practice with local dairy cows, which graze in open fields and have to search for water and grass. Improved dairy cows have also shifted gender division of labour and their associated benefits from male to female. Keeping Boran x Jersey dairy cows was financially profitable and their products had nutritional benefits, enhancing household. Access to a reliable artificial insemination service and forage throughout the year continue to be major challenges in smallholder crossbreed dairy farming at Melkassa.

Keywords: Farmer Research Group, gender, smallholder farmers, dairy, Boran x Jersey breed, Melkassa

Introduction

Ethiopia has the largest livestock population in Africa of which cattle number some 55.48 million (CSA, 2013). Cattle provide food, cash income, draft power, manure for soil fertility improvement, employment opportunities and raw material for industries. For instance, livestock contributes 12-16% of national gross domestic product (GDP) and 30-35% of agricultural GDP, where 60-70% of the population are engaged in agriculture (SVN, 2008).

Dairy farming makes a significant contribution to the food and nutritional security of farm household. The performance of Ethiopian local breed dairy cows is, however, low in East Africa compared for example with Kenya, Uganda and Tanzania and with international benchmarks. The average

http://dx.doi.org/10.3362/9781780449005.008

daily milk yield per cow is only 1.8 litres, lactation duration is short (5.9-9.8 months), the first calving age is late (53 months), and the interval between calves is long (21 to 25 months) (Tegegne et al., 2013; SNV, 2008). The present world average milk yield is 25 litres per day, lactation length is 10 months and average calving interval is 15 months. In Ethiopia, Dinka (2012) reported from his study in the Assela area that crossbreed cows have their first calves at 35 months and then produce calves at 12.4 month intervals. Fortunately, there are both demand side and supply side drivers for dairy sector development in Ethiopia. The demand side drivers are rapid population growth[1], opportunities for employment, a present and growing shortage of dairy products in major cities and the dependency of the country on imported milk products (SVN, 2008). The supply side drivers include better breeding technologies, improved forage production and cheaper energy, all of which will positively affect the supply of dairy products.

Smallholder dairy development provides opportunities to address the persistent problem of poverty in rural and peri-urban areas. The marketing of dairy products transfers income from affluent urban households to their poorer rural counterparts (Moran, 2009; FAO, 2008). The dairy sector also improves food and nutritional security for both rural and urban households. Dairy development usually enhances the living standards of traditional smallholder farmers and dairy producers, traders (including wholesalers and retailers), distributers and consumers, and improves the nutrition of poor households, mainly by improving infant nutrition. Additionally, the dairy sector sustains the natural resource base by recycling dairy farm by-products for crop production.

Dairy development is associated with technical changes to increase milk yield per cow and also with improvements in dairy cow management. Using highly productive cows[2] is a rapid and potentially sustainable path to higher productivity, including for resource-poor farmers living in hot, semi-arid climates such as the Central Rift Valley and many other parts of Ethiopia. Dairy development involving more productive cattle has to be integrated with fodder crop technology, since raising dairy yield is the product of both genetic potential and good management (Ouma et al., 2007). Since technical changes are required alongside financial resources, the participatory Farmer Research Group approach is ideal.

In developing nations, such as Ethiopia, dairy farming is mainly based on open grazing. Cattle management is largely taken care of by men and boys. However, as a result of introducing improved dairy cows, there are often increased expectations of a shift in the division of labour, benefits and resource-sharing between men and women in smallholder dairy farming.

For the purpose of understanding and analysis, a dairy farm is categorized based on the size and types of management, based on Moran (2009). He categorized dairy farm into smallholder (those farmers who keep less than 20 heads of milking cows); semi-commercial households (those that own 20 to 50 heads of milking cows); and commercial (dairy farmers who own more than 50 milking cows). In all the categories, the farmers have to keep replacement heifers. This definition is adopted in this chapter because it is applicable in the context of the study area where smallholders are predominant.

Although there is consensus on the economic and nutritional benefits that result from keeping improved dairy cows, smallholder farmers face challenges in their access to improved cattle owing to a shortage of supply and the need for intensive management practices (i.e., increased feed requirement and the need to maintain the blood levels of improved dairy cows), which in turn requires new knowledge, skills and financial resources. Likewise, the shift from local dairy cows to crossbreed cows is expected to be accompanied by a shift in gender roles and gender division of labour.

Description of the study area and method of data collection

The study was conducted in the Melkassa area in Wakie-Mia-Tiyo *Kebele* in Adama district, East Shewa zone, and in Awash-Bishola *Kebele*, Doddota district, Arsi zone. The main source of livelihood is agriculture, with a significant proportion of the population also involved in small-scale trading, casual labouring and the civil service. A household on average constituted 5.94 persons equally made up of male and female members. Female-headed households constitute 17.7% of the total. The age dependency ratio (those in the age group below 15 and above 64 compared with those in age group 15 to 64 years) was high at 63.1%. In terms of religious affiliation, Christianity dominates with Orthodox (54.5%), Protestant (32.2%) and Catholic (0.7%), while the remaining population (12.6%) divides into Muslim and *Waqefata* (Table 8.1).

Livestock populations and land use

The most frequently farmed livestock are cattle, then sheep and goats, donkeys and horses. Most cattle are local breeds, except for a few introductions by the FRG project and a few resulting from artificial insemination (AI). The land is mainly used for farmland (both rainfed and irrigated), houses and other local services. In Wakie-Mia-Tiyo, sugar cane and food crops occupy 57.32% of the land while the remaining is used for local services (school, clinic, administrative offices, centres, farmer training centre, etc.). There was no grazing land in the *Kebele* of the 2873.76 haectares. In Awash Bishola *Kebele*, in Bishola sub *Kebele*, farmland occupies 56.69%, grazing land 9%, and the remaining are residential areas from the 837 ha.

Method of data collection

Data were obtained from 12 farm households in the Melkassa area that had participated in crossbreed dairy cow FRG activities since 2007. Data were collected by the author and his assistant in May 2014, using questionnaires. The information included dairy production, housing, feed sources, health and

Table 8.1 Population of the study *Kebeles* by gender, age and religion

	Wakie-Mia-Tiyo	Awash-Bishola	Total	%
Population (total)	4,721	6,784	11,505	100.0
Male	2,380	3,463	5,843	50.8
Female	2,341	3,321	5,662	49.2
Household head (total)	841	1,094	1,935	100.0
Male	716	876	1,592	82.3
Female	125	218	343	17.7
Under 15 years	1,643	2,610	4,253	37.0
15-64 years	3,064	3,989	7,053	61.3
Above 64 years	14	185	199	1.7
Christian	4,485	5,571	10,056	87.4
Orthodox	944	5,327	6,271	54.5
Protestant	3,541	168	3,709	32.2
Catholic	-	76	76	0.7
Muslim	-	1,069	1,069	9.3
Waqefeta	236	144	380	3.3

record keeping. Likewise, group discussions were conducted with the dairy farmers and experts involved in the process.

Crossbreed dairy heifers (Boran x Jersey) were introduced to the farmers from Adami Tulu Agricultural Research Centre. Boran cattle is an Ethiopian breed known for its higher milk yield, higher fat content milk, and better disease resistance compared with other local breeds (Dr Birhanu Shelema, personal communication, 2014). The high fat content of Boran milk enables smallholders to produce butter during times of lower demand for fresh milk (e.g. during Orthodox Christian fasting periods, or because they are a long way from market). Similarly, Boran breeds perform well in semi-arid areas in times of feed and water shortages. The Jersey[3] breed is an exotic cattle breed in Ethiopia. Jersey milk has a higher butter content compared with other exotic breeds such as Holstein Friesian, and they give a higher milk yield than local breeds. Hence Boran x Jersey cattle were anticipated to fit well into smallholder farm conditions in the area, while helping to minimize post-harvest losses by producing rich milk suitable for making butter.

Site and farmer selection

Wakie-Mia-Tiyo in Adama district and Awash Bishola in Doddota district were selected for their dairy Farmer Research Groups. The sites had been initially selected for vegetable FRGs. The farmers in these areas are essentially vegetable producers, using irrigation water from the Awash River. The selection of farmers for dairy research was carried out based on their interest and requests following their visit to Debre Zeit modern vegetable farm, which is also a dairy farm. The request was picked up by Melkassa Agricultural Research Centre because of the level of interest in dairy shown by the group. The farmers had requested the opportunity to access, learn about and use improved dairy cows, and utilize available forage such as sugarcane and food factory by-products and also explore the possibility for forage production. The criteria for selecting the sites and farmers included ease of transport and communication, and proximity to potential markets for milk in fast-growing urban centres such as Awash Melkassa and Adama. The potential to apply cow dung as manure in vegetable production was also a consideration in farmer and site selection.

The selection of individual farmers also took into account their willingness to assign part of their land for forage production and to try new feed management options, as well as their ability to contribute the down payment and balance at the agreed times. Their willingness to build shade for the heifers, to keep records (in the agreed format), as well as readiness to share their experiences with surrounding farmers and visitors from other areas were also factors in the selection process (Endeshaw et al., 2009). Roles and responsibilities for the major actors were identified and shared (Table 8.2).

The structure of the dairy FRG was created within the existing and larger vegetable FRG. The dairy FRG members cooperated with each other to arrange the necessary training, the management of revolving funds (a source of money from which loans are made for improved dairy farmers to benefit a larger number of farmers), hosting field visits and organizing field days to share experience with other farmers.

Results and discussion

Description of the households

The households selected fairly represent the wider farming population of the area. One-quarter of the interviewed households were female-headed; family sizes were about average at 2.4 adult male and 2.7 adult female members. Among the respondents, four were not able to read and write, two of them could read and write[6] and the remaining six farmers had an average of 7.7 (standard deviation[7] = 2.33) years of schooling[4]. The respondents were adults of average 48.8 (SD=9.7) years. Nine respondents were married and living with their spouses, while three were widows. Ten of the respondents were Christians (nine Orthodox) while two of them were Muslim.

The issue of gender and division of labour becomes more important since the improved dairy management is more intensive than with local breeds. Hence, a closer look at household gender and division of labour is vital. An average household is composed of 7.7 (SD=3.6) persons (Table 8.3). The respondents mainly comprised nuclear families of a husband, wife and children, though

Table 8.2 The roles of partners in improved dairy cow introduction and management

Farmers	Researchers	Agricultural experts, subject matter specialists and development agents
• Purchase Boran x Jersey heifers and carry out recommended management practices • communicate problems or developments to extension agents/ researchers • Share information with other farmers and visitors • Keep records on blood level between the cross breed, disease episodes, milk yield, forage/feed used • Grow and manage forage crops • Prepare silage for dry season	• Deliver pregnant Boran x Jersey heifers • Prepare initial data recording sheet and provide introductory information to farmers • Train farmers on materials/mechanisms required in measuring milk and feed • Regular monitoring of dairy management practices • Train farmers and extension staff in required skills (e.g. silage making)	• Monitor and facilitate smooth implementation of activities • provide AI and vet services • Facilitate farmer-to-farmer experience sharing • Supervise forage cultivation and post-harvest management practices • Manage revolving funds including overseeing credit repayments

Note: Extension workers at the village level are called development agents (DAs) in Ethiopia. Subject matter specialists serve as extension workers providing training and technical backstopping to DAs.
Source: Modified from Endeshaw et al., 2009

Table 8.3 Improved dairy farm household composition by age and gender (n=12)

Age category	Male (Average)	Female (Average)	Total (Average)
Less than 7 years	0.3 (0.5)	0.1 (0.3)	0.4 (0.5)
7–14 years	0.8 (0.7)	0.9 (0.8)	1.6 (1.4)
15–64 years	2.4 (1.2)	2.7 (2.5)	5.1 (3.4)
> 64 years	0.2 (0.4)	0.2 (0.4)	0.3 (0.7)
Total	3.7 (1.4)	4.0 (2.5)	7.7 (3.6)

there were a few households consisting of extended family of both of the spouses' parents or their own grandchildren. The age dependency[8] ratio was 0.45 where one adult supports one child or an old-age family member. In this respect, women provide essential household labour in taking care of dependaents and sick family members, because women are responsible for almost all the caring activities of a household.

The farmers mainly practise mixed farming. Horticultural (vegetables and fruits) crop production is undertaken using furrow irrigation, while cereals and pulses are rain-fed. The cereal and pulse crops produced include maize, teff, common beans, wheat and barley. The average household landholding is 2.1 ha, where the rain-fed areas constitute 0.58 ha. A household allocates about a quarter of a hectare for irrigated crop production (Table 8.4).

Table 8.4 Average land ownership and land allocation of improved dairy farmers (n=12)

Area (ha)	Mean	SD
Rain-fed land	0.58	0.94
Irrigated land	1.50	1.99
Crop area		
Maize	0.16	0.19
Teff	0.25	0.46
Common beans	0.04	0.14
Wheat	0.04	0.14
Barley	0.02	0.08
Onion	0.45	0.91
Tomato	0.19	0.18
Sugar cane	1.19	1.96
Pepper	0.04	0.10
Fruits	0.04	0.10
Total landholding (ha)	2.10	1.99

Table 8.5 Types of livestock kept by respondents (n=12)

Type of livestock	Average number of livestock owned	
	Local	Cross-breed
Oxen	1.17 (1.27)	0.08 (0.29)
Cows	0.42 (0.67)	1.42 (1.08)
Heifers	0.42 (0.67)	0.75 (0.75)
Bulls	0.08 (0.29)	0.17 (0.39)
Calves	0.33 (0.65)	1.00 (1.04)

Note: Figures in brackets indicate the standard deviation

Livestock production

Livestock production is important for farm households in the area. The most frequently kept livestock are cows and oxen. The most commonly kept dairy cows are local breeds. The number of locally bred cows is, however declining rapidly and their productivity is low. Among the respondents, only one-third indicated that they owned a local breed – mainly oxen for land preparation. Locally bred oxen are the major sources of draft power, while improved dairy cows have become important sources of milk among the respondents (Table 8.5).

The lactation period for both improved dairy cows and local breeds was nine months. The farmers indicated that local breeds usually have a shorter lactation period and their milk yield also declines more rapidly. The average milk yield of a local cow was 2.1 litres/day while that of a crossbred cow was 7.8 l/day. The milk yield of crossbred was 3.6 times higher than that of the local dairy cows and the difference tends to become greater as lactation time increases (Table 8.6). The most striking difference between local and improved cows is observed in their first calving. The first calving was at 31.6 months for the crossbreed, while it is 46 months for local breeds. The local breeds also require a longer calving interval of 19.4 months (average) as compared to a 12.6 month calving interval for local breeds.

Gender division of labour in improved dairy farming

As shown in Tables 8.7 and 8.8, the labour contribution of women in male-headed households is higher than any other member of the household in dairy cow management. This is because women's involvement in most of dairy cow management takes place at home, demanding much more of women's labour and time. The picture is different for local dairy cows. In local breed livestock farming men and boys have the key roles in looking after livestock away from home, searching for pasture and water, accessing information on breeding, animal health and cow heat (readiness for mating) detection. Whereas in local dairy management, women are involved in activities based at home, such as cleaning, milking and milk processing (Tegegne et al., 2013). The introduction of improved dairy cows has changed the gender division of labour by shifting some activities from men and boys on to women and girls.

Improved dairy cows increase the demand for women's labour, but there is a commensurate increase in the economic benefit to women since the revenue accrued from selling milk and butter is controlled by women. Women have got more power in decision-making in livestock sales and the allocation of income obtained from them. Moreover, women farmers have become involved in the training and exchange visits to the same extent as male farmers (Tables 8.7 and 8.8).

Table 8.6 Lactation period and average milk yields of local and crossbred cows

Lactation period (months)	Average milk yield (litre/day) of cows		
	Local (n=4)	Crossbreed (n=12)	Milk yield ratio (crossbreed cows to local)
Early lactation (< 3)	2.88 (2.18)	10.42 (3.83)	3.61:1
Mid-lactation (3–6)	2.00 (2.04)	7.20 (3.09)	3.60:1
Late lactation (6–9)	1.50 (1.68)	5.81 (1.84)	3.87:1
Total	2.12 (1.95)	7.81 (2.71)	3.68:1

IMPROVED DAIRY PRODUCTION AND CHANGING GENDER ROLES   125

Table 8.7 Gender role-sharing in livestock management in male-headed households (n=9)

Dairy management	Men	Women	Boy(s)	Girl(s)
Feeding	6	13	6	6
Providing water	3	9	8	8
Cleaning shelter	2	15	7	5
Milking	0	18	0	1
Selling milk	2	17	0	0
Making butter	0	18	0	2
Heat detection or observation	7	11	3	1
Training participation	9	10	0	0
Live animal sale	11	10	0	0
Total score	4.4	13.4	2.7	2.6

Note: The participation of farm household members in a particular activity was scored as 'usually' (2), 'sometimes' (1) and 'no' (0). Accordingly, each able household member can make a maximum contribution equivalent to a maximum score of 18 points, that is, if certain household members usually handle all the nine activities, say the woman (wife).

Table 8.8 Gender role-sharing in livestock management in female-headed households (n=3)

Dairy management	Women	Boy(s)	Girl(s)
Feeding	4	1	2
Providing water	4	1	3
Cleaning shelter	5	0	3
Milking	6	0	0
Selling the milk	6	0	0
Making butter	6	0	2
Heat detection or observation	6	2	1
Training participation	6	0	0
Live animal sale	4	0	0
Total score	15.7	1.3	3.7

Housing for dairy cows

Housing is essential to protect the cows from wind, sun and dangerous wild animals. Keeping cows in shelters also facilitates easier feeding and watering. Six (half) of the study participants have built houses that protect the dairy cows from sun and wind. Four farmers built shelters which protect the cows from sun and wind, and also have sewage removal. Two households continued to use the traditional *kraal*, with no protection from sun and wind. In general, the dairy FRG members have better housing for their dairy cows compared with local dairy cattle owners, since almost all of the latter

Figure 8.1 A dairy FRG member feeding her Boran x Jersey heifer in open space in Wakie-Mia-Tiyo, May 2014

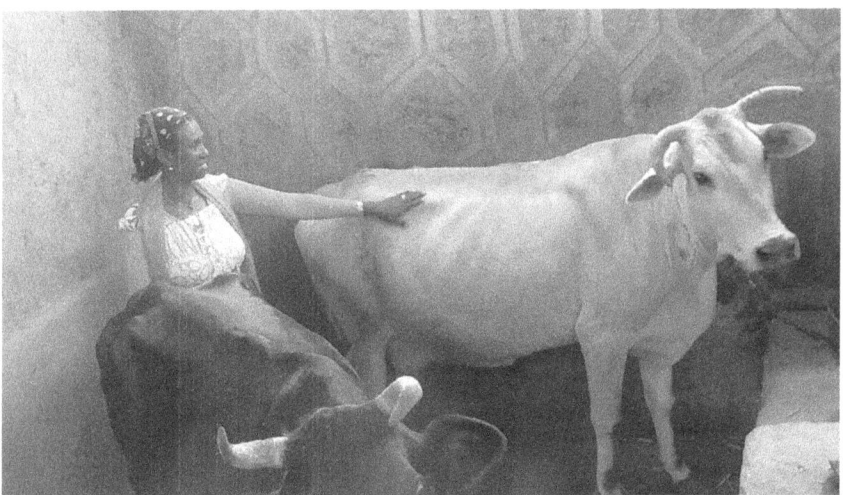

Figure 8.2 A dairy FRG member caring for her Boran Jersey cows in a concrete built house, Awash Bishola, May 2014

use open pens without roofs which are not adequate to protect the dairy cows from sun and wind. This is reflected in the lower milk yields. The dairy FRG farmers use local oxen for ground preparation, but they send them to more rural, open areas for grazing, as these local breeds can survive on less feed as compared to the introduced crossbreeds.

Livestock feed and feed management

The farmers mainly feed their dairy cows on sugar cane, teff straw, maize stalks, and various concentrates of wheat and maize products. The FRG member farmers were also trained in improved forage production, using fresh maize stalks for their animals during the dry season when there are feed shortages. Almost all the dairy farmers are now familiar with improved forage preparation methods, such as silage making. They were given forage crops such as pigeon peas and elephant grass to grow. However, the farmers were not practising silage making and growing forage crops as expected at the time of this study. The farmers indicated that land shortages were the major constraint for forage crop production. Slightly more than half (seven out of 12) of the farmers pointed out that feed shortage was becoming a challenge, as the price of concentrates was rising at an alarming rate[5].

Zero grazing is a system where animals feed on cut grass in the shelter instead of putting them in fields to graze. Zero grazing has higher financial advantages, coupled with access to collective marketing and credit (Wambugu et al., 2011). It is also claimed as optimal utilization of land owing to its higher output per unit area. It also reduces pests (e.g. ticks and worms) and damage to crops (Ouma et al., 2007). Zero grazing also minimizes environmental damage caused by over-grazing and the trampling effect which exposes bare soil to erosion.

Livestock health and artificial insemination services

Livestock health is generally good and disease is not a serious problem in the area. When cattle get sick, famers call or visit the nearby public or private veterinary clinics[6]. Likewise, some farmers treat sick animals by themselves or invite traditional cattle healers. Almost all the farmers receive veterinary services at their residences/villages. Farmers indicated that the common diseases in the area are those easily controlled by local treatment. Moreover, the improved livestock are not as susceptible to those diseases.

The issue is rather obtaining reliable artificial insemination services (particularly for Wake-Mia-Tiyo farmers). Veterinary services (including artificial insemination) are provided by the district office of agricultural development. Veterinary technicians come from Adama for Wakie-Miya-Tiyo farmers and from Dhera for Awash-Bishola farmers. Adama is located 18 km away while Dhera is located 5 km from the farms. It seems that the distance travelled has a strong influence on the AI service. For instance, in Awash-Bishola most of the calves produced are a result of artificial insemination while in Wakie-Mia-Tiyo almost all of the calves born are from live animals of the local breed or from a live animal of unknown origin. AI is claimed to be problematic in this area in terms of the need for repeated services. In this study, the service is carried out an average number of three times (2.73, SD=1.6) with a range of two to six times.

Record keeping

Record keeping was an important issue to cover in the farmers' training. Most of the farmers told us that they have now been trained to keep records covering the quantity and cost of feed (expenditure), income, and the health of the cattle (Table 8.9). There was a direct association between record keeping and the amount of formal education received by the household heads. Households headed by women and the illiterate were observed to be relatively poor at record keeping, even though they have literate children of an age capable of helping.

Table 8.9 presents the costs and benefits of local and Boran x Jersey cows. Table 8.10 summarizes the synthesis.

Table 8.9 Costs and benefits of keeping crossbreed (Boran x Jersey) dairy cattle

Item	Milk (litres)	Revenue (birr)	Estimated income (birr)
Milk sold average no. of			
a. first 3 months	9.42	870.30	8,198.20
b. 3-6 months	7.20	870.30	6,266.20
c. 6-9 months	4.81	870.30	4,186.10
Total income per lactation per cow			14,205.36
Annual feed cost per animal		18,435.90	
Net income per lactation per cow			4,230.54

Note: Revenue equals the sales of milk over three months (90 days x 9.67 birr per litre).

Conclusions and recommendation

Dairy farming is a traditional agricultural practice in Ethiopia. The productivity of local dairy cows is low and grazing lands are shrinking at an alarming rate, or absent altogether in many villages in the Central Rift Valley. The introduction of crossbreed cows has been found to be a promising option. Boran x Jersey heifers were introduced to groups of farmers in Melkassa. Since no comparison was made between FRG and non-FRG members, it is not possible to state definitively that the FRG was essential for learning new techniques, or for enhancing payments of the revolving fund. However, by observing the continuation of activities and adoption of technologies among ex-FRG member farmers, as well as interactions between researchers and ex-FRG members, the FRG is certainly a very important element in such improvements.

Crossbreed cows need different labour contributions by household members. In this study, the contribution of women in both male-headed households and in female-headed households has increased. That is because crossbreed cattle

Table 8.10 Synthesis of the comparison of local breed and crossbreed dairy cows in the study area

Descriptor	Local breed dairy cows	Crossbreed dairy cows
Reproductive performance	Have later first calving (46 months), longer calving interval (19.4 months) and lower average milk yield (~ 2.1 litres/day).	Have earlier first calving (31 months), shorter calving interval (12.6 months) and higher average milk yield (~ 7.8 l/day).
Performance under hardship	Local breeds are foraging in open fields. They survive on low amount of feed with minimum productivity. There is a continual feed shortage, except during the rainy season and short harvest season when they feed on crop residues.	Since feed is planned, feed shortages are not an important issue. However, the price of feed is currently growing at an alarming rate.
Environmental issues	Grazing in fields can damage farmland (and roads), leaving the soil exposed to erosion.	Being kept in or around the house minimizes erosion caused by widespread grazing.
Contribution to household food	The milk yield is too low to meet the need for all household members. The number of local-breed cows is declining.	Give high yield and all households are able to access milk from their own cows. The number of crossbreed cows is increasing.
Management	Local breeds are adapted to open grazing, which causes land to degrade. This means the amount of available grazing land is shrinking at an alarming rate.	Crossbreed cattle are mostly kept in shelters, which avoids land degradation. Moreover, the waste by-products can easily be collected and used as fertilizer.
Labour demand	Local breeds need herding, usually attended to by children (boys) throughout the day.	The labour required is spread throughout those of working age in a household. This releases school-age children for studying, since animals are kept close to the home.
Suitability for breeding	Fertilization is by live bulls. There are difficulties getting bulls of the desired blood level since the process is random, making breed improvement very challenging.	Convenient for AI service since cows are kept in a controlled environment. The present challenge is getting AI services on time.
Gender division of labour	Women (and girls) are involved in milking and cleaning houses or *kraal*. Children (mainly boys) are main labour source for herding cows in the field and taking to water points.	Women take care of feeding, providing water and cleaning shelter. Men also share these activities to some extent. Responsibilities of children are reduced. Milking and selling milk is still the sole responsibility of women. Decisions about selling animals and sharing benefits is jointly made between spouses.

are kept and fed in shelters, unlike the local cows. The gender labour demand of dairy management overall has shifted from male (men and boys) to female (women and girls).

The environmental damage caused by trampling cattle herds is also minimized because the crossbreed dairy cows are kept under shade near the house.

Finally, the income and milk consumption of the farm household has increased through the introduction of crossbreed cows, contributing positively to smallholder farm households' food security and nutritional intake.

Acknowledgements

I would like to thank the crossbreed dairy Farmer Research Groups in Awash-Bishola and Wake-Mia-Tiyo *Kebele*s and the respective agricultural development workers for their time and assistance in the fieldwork. I am grateful to Gadissa Ejersa for his assistance with data collection.

About the author

Bedru Beshir has been affiliated with the Ethiopian Institute of Agricultural Research at Melkassa Agricultural Research Centre since June 1999. He studied regional and local development for his Master's degree at Addis Ababa University and carried out his PhD in International Development at Nagoya University, Japan. He has worked on crop and livestock technology promotion and dissemination as a researcher and department head of the centre's outreach programme.

Notes

1. The current population growth of Ethiopia is 2.89% while the population is estimated at 96.6 million (World Fact Book, 2014). Moreover, the urban population is growing more rapidly and may be able to afford the price for packaged milk.
2. Cows which have higher milk yields, shorter period until the first calving and a shorter calving interval, efficiently use forage, produce better meat and are suitable for intensive farming.
3. The Jersey cow is relatively low in weight at 400–500 kg. Its lower body weight means it has lower maintenance requirements, superior grazing ability, high fertility, high butterfat content, and can thrive on locally produced forage. Bulls are also small, ranging from 540 to 820 kg <http://en.wikipedia.org/wiki/Jersey cattle>.
4. These reading and writing skills were obtained from informal (basic) education provided during the Dergue Military Rule regime (1974-1991 under the leadership of Mengistu Haile Mariam). As farmers have commented, such skills are essential yet lacking in their area.
5. The average price of 1 kg concentrate of *fagulo* (factory by-product from a processed crop) was ETB 3.50 in 2012, 4.75 in 2013 and 10 in 2014, while the price of 1 kg concentrate of *furushka* (factory by-product from

processed crop of different crop combination and quality to *fagulo*) was ETB 1.37 in 2012, 2 in 2013 and 4.50 in 2014. From 2013 to 2014 (July), the concentrate price increased by 100%.
6. There are private veterinary clinics nearby in Awash-Melkassa town and public clinics in the main towns of each district, Dhera and Adama.

References

Aregu, L., Bishop-Sambrook C, Puskur R and Tessema E. (2010) *Opportunities for promoting gender equality in rural Ethiopia through the commercialization of agriculture*, Nairobi: IPMS (Improving production and market success) Ethiopian Farmers Project Working Paper 18. IRLI (International Livestock Research Institute). Addis Ababa.

CSA (Central Statistical Agency) (2009) Agricultural Sample Survey 2012/2013 (2005 Ethiopian Calendar). Report on Livestock and Livestock Characteristics, *Statistical Bulletin 570*, Volume III, Addis Ababa.

Dinka, H. (2013) 'Reproductive performance of crossbred dairy cows under smallholder conditions in Ethiopia', *International Journal of Livestock Production*, 3(3), pp. 25-28. <http://dx.doi.org/10.5897/IJLP11.055>.

Endeshaw H., Aklilu M., Chali Y., Chimdo A. and Belay D. (2009) 'Introduction of Feed Crop Production and Processing in Irrigated Vegetable Farming System in Adama and Dodota Districts'. In: Bedru B., Wole K., Niioka M. and Shiratori, K. *FRG completed Research Reports*, pp. 180-191. EIAR, OARI and JICA, Addis Ababa.

FAO (2008) *Asia Smallholder Dairy Development Strategy and Outline Investment Plan*, 'Improved Market Access and Smallholder Dairy Farmer Participation for Sustainable Dairy Development', Common Fund for Commodities Animal Production and Health Commission for Asia and the Pacific Food and Agriculture Organization of the United Nations, Bangkok. <www.fao.org/ag/againfo/themes/documents/dairy_dev_strat.pdf>

FAO (2011) *Successes and Failures with animal nutrition practices and technologies in Developing countries*, Proceedings of the FAO Electronic Conference, 1-30 September 2010, Rome. Edited by

Metaferia, F., Cherenet, T., Gelan, A., Abnet, F., Tesfay, A., Abdi, J., and Gulilat, W. (2008) *A Review to Improve Estimation of Livestock Contribution to the National GDP*, Ministry of Finance and Economic Development and Ministry of Agriculture, Addis Ababa p. 42. <https://cgspace.cgiar.org/bitstream/handle/10568/24987/IGAD_LPI_GDP.pdf?sequence>.

Moran, J. (2009) *Business Management for Tropical Dairy Farmers*, CSIRO, Canberra, Australia. <http://www.publish.csiro.au>.

Negassa, A. (2009) Improving smallholder farmers' market supply and market access for dairy products in Arsi Zone, Ethiopia. Research Report 21. International Livestock Research Institute, Nairobi.

Ouma, R., Njoroge, L., Romney, D., Ochungo, P., Staal, S., and Baltenweck, I. (2007) Targeting dairy interventions in Kenya: a guide for development planners, researchers and extension workers. International Livestock Research Institute Manuals and Guides No. 1. Nairobi.

SNV Netherlands Development Organization (2008) *Study on Dairy Investment Opportunities in Ethiopia*. Addis Ababa.

Tegegne, A., Gebremedhin, B., Hoekstra, D., Belay, B., and Mekasha, Y. (2013) *Smallholder dairy production and marketing systems in Ethiopia: IPMS experiences and Opportunities for market oriented development*, Nairobi: IRLI, IPMS (Improving production and market success) Ethiopian Farmers Project Working Paper 31.

Wambugu, S., Kirimi, L. and Opiyo, J. (2011) Productivity Trends and Performance of Dairy Farming in Kenya, Nairobi, Tegemeo Institute of Agricultural Policy and Development.

CHAPTER 9
Farmers' perceived benefits of FRG-based research: the case of selected FRG-based research activities

Shingo Takeda

Abstract

A programme of gathering qualitative data took place, on the benefits perceived by farmers through Farmer Research Group (FRG) research activities, based on the farmers' own experiences in adopting the technology, their reasons for the adoption, the farming challenges, and the direct and indirect benefits. The results revealed that a majority of the farmers reported positive benefits with one-third reporting successful adoption of the technologies targeted by the research and only a third of the farming challenges were covered in the research our project supported. Nearly half of the farmers learned some new skills and processes during the project and incorporated them into their farming practices. The farmers evaluated and adopted useful technologies based on various aspects rather than on a single criterion. These findings indicate that the FRG-based research contributed positively to farming practices. Matching the farmer's needs on farming with subject to be addressed in the research is important and requires improved communication between farmers and researchers.

Keywords: farmers' perception, benefit, technology adoption, farming challenges, Farmer Research Group, Ethiopia

Introduction

The Farmer Research Group (FRG) approach is a participatory research approach that aims to develop technologies appropriate for various natural and socio-economic environments through the synergy created between researchers' scientific knowledge and farmers' indigenous knowledge (see Chapter 3 of this book: Seo, 2016). The Project for Enhancing the Development and Dissemination of Agricultural Innovations through Farmer Research Groups (FRG II) promoted and expanded the FRG approach through training, supporting FRG-based research activities and enhancing the communication of research outputs through developing extension materials within the Ethiopian national agricultural research system (NARS). The FRG-based research activities were conducted by a number of NARS researchers, who had received training on the FRG approach and on developing extension materials. The FRG-based research activities were carried out by various stakeholders, with farmers and researchers being the main actors.

Precise socio-economic assessment is needed to assess the overall impact of the research; however, this can only be carried out after the project has concluded. Therefore, an assessment of farmers' perceptions was conducted with researchers during the project, to assess the extent of technology adoption, the circumstances influencing the adoption, the direct and indirect benefits, and the challenges farmers face. This assessment was carried out through holding interviews with both FRG member farmers and non-FRG farmers, focusing on the perceptions that prompted the farmers to behave as they did. The result of that assessment is discussed in terms of the FRG approach.

Methods

Targeted research

Out of the 43 research projects that FRG II supported during the project period between 2010 and 2015, 13 were selected for this assessment as they had been implemented before 2013 and had been in progress for at least one year at the time this study took place in 2014. All the selected research activities were crop-related, which allowed for two assessments to be made over consecutive production seasons (Table 9.1).

Interviewed farmers

Instead of a random sample taken from the whole population of participating farmers – the usual method for evaluating the impact of an intervention – farmers who were able to take part in the interviews were selected based on the assessment objective: to extract qualitative data with the aim of further improving FRG-based research activities. Sixty-one farmers from the village where the FRG-based research took place who could take part on the interview date were selected and interviewed in the field. A semi-structured interview was conducted with interpretation by and support of researchers. The farmers who had participated in the FRG research (FRG member farmers) were the priority for the interview; however, when non-FRG farmers were available they were also included for comparison. Out of the 61 farmers interviewed, 13 were non-FRG farmers (Table 9.1).

Interview procedure. The interviews were conducted with individuals one at a time, to avoid influence from others. The interview was structured as simply as possible considering how busy the farmers were, focusing on the perceptions that prompted changes in their farming practice. The contents of the interviews and their objectives are shown in Table 9.2.

Table 9.1 Research targeted for assessment, date of interview, and numbers of interviewed farmers, 2014

	Organization/ university	Title of the research	Season(s) covered by research	Date of the interview	No. of interviewed farmers	
					FRG	Non-FRG
1	Fedis ARC	Participatory evaluation of difference rates of fungicides (Mancozeb and Apron plus 50%) for the control of ground root rot (Sclerotium Ralfsii Sacc) in Eastern Harage.	2011	10.6.2014	4	0
2	Adet ARC	Demonstration of rice transplanting and seed pre-germination technology in Fogera Plan.	2010 2011 2012	2.7.2014	2	0
3	Adet ARC	Participatory variety selection of rice with farmers in rain-fed conditions.	2011 2012	3.7.2014	3	0
4	Bahir Dar University	Seed multiplication rate study of different cultivars of wheat for optimizing seed quality and productivity.	2011 2012	14.7.2014	4	0
5	Bahir Dar AMFSRC	Participatory evaluation and demonstration of animal-drawn compactor.	2011 2012	15.7.2014	4	0
6	Bako ARC	Evaluation and promotion of improved teff technologies in two selected districts: East Wollega and Horo Guduru Zones.	2012 2013	5.8.2014	6	0
7	Sekota ARC	Participatory evaluation of seed-dressing insecticides for the management of teff shoot fly, Wag-Lasta area.	2011	13.8.2014	4	3
8	Mekelle University	Participatory evaluation of agronomic performance of farmer-saved and purified wheat seed.	2011 2012 2013	21.8.2014	4	2
9	Wolaita Sodo University	Participatory evaluation of lower teff seeding rates using seed spreaders in Wolaita Zone.	2010 2011 2012	29.8.2014	4	1

(continued)

Table 9.1 List of research targeted for assessment, date of interview, and numbers of interviewed farmers, 2014 (*continued*)

	Organization/ university	Title of the research	Season(s) covered by research	Date of the interview	No. of interviewed farmers	
					FRG	Non-FRG
10	Adami Tulu ARC	Effect of seed treatment on seed-borne diseases of hot pepper and onion.	2011 2012	6.11.2014	5	0
11	Werer ARC	Participatory rice seed rate determination for broadcast sowing under irrigated ecosystems.	2010 2011 2012 2013	5.12.2014	2	3
12	May-Tsuberi ARC	Enhancing production and productivity of upland rice through quality seed production in North Western Tigray, May-Tsuberi District.	2011 2012	17.12.2014	3	2
13	Wolaita Sodo University	Participatory evaluation of seed treatment techniques to improve tomato seed quality.	2011 2012 2013	14.1.2015	3	2
Total					48	13

ARC: Agricultural Research Centre, AMFSRC: Agricultural Mechanization and Food Science Research Centre

Table 9.2 Contents of the interview and objectives

	Contents of the interview	Objectives
1	Changes in yield and cultivated land area of target crop and reason(s)	To know if research process or technology contributed to increase in productivity and/or area under cultivation
2	Adoption of the technology and reason(s)	To identify the outcome of the research and understand criteria for adoption
3	Challenges encountered in cultivation of the target crop	To identify major farmers' problems in cultivation and to make sure if the problems were addressed in the research
4	Benefits obtained from the research	To confirm reason(s) for changes in yield and cultivated land area of target crop, and whether benefits were created by the technology or by other factors

Note: Multiple answers were allowed for parts 3 and 4

Text-mining and quantification

Text-mining and quantification were performed according to Sato (2008 : 211). The grouping exercise mentioned below was determined by reference to Jones

(2001) and Cately et al. (2007). In the text-mining process, the most frequent similar responses of farmers are grouped first and the farming attribute is considered for the naming of the groups.

Changes in yield and cultivated land area of targeted crop. Specific figures for changes in yield and cultivated land area were placed in one of three categories: 1) Increased; 2) Decreased; 3) No change. The matching of reasons for 1) with 'farmer's benefit obtained from the research' was counted to understand whether the research had contributed to an increase in production or the area under cultivation. The frequency of a match and its percentage were tallied.

Farmers' adoption of the technology. The total number of adopters was counted and the ratio of the adopter to total number of interviewees was tallied for each of the FRG member farmers and non-FRG farmers. Each reason for the adoption was categorized into one of the following seven groups: 1) Labour-/cost-/time-effective; 2) Higher yield, 3) Higher market price; 4) Superior agronomic performance (observed as morphological or physiological feature of the plant); 5) Superior tolerance/resistance; 6) Superior cooking ingredient; 7) Other (seed availability, simplicity of use, etc.). Each reason for non-adoption was also categorized into one of the following six groups: 1)The technology is not suitable for farming environment/condition; 2) There is a suitable/better alternative; 3) Not remembering the technology; 4) Requires more labour/time; 5) No access to the technology; 6) Other (lack of knowledge, inputs, etc.). The frequency of replies in each category was tallied.

Challenges in cultivating the target crop. Each challenge was categorized into one of the following ten groups: 1) Cost problems; 2) Problems related to agronomic practice 3) Diseases; 4) Pests; 5) Problems relating to land; 6) Problems related to inputs; 7) Problems related to water; 8) Problem related to labour; 9) Market problems; 10) Other (no challenge, climate etc.). The frequency of replies in each category was tallied. Additionally, challenges covered in the research were counted and its ratio to total number of the answered challenges was tallied in order to understand how much researchers' respected farmers' needs.

Farmers' perceptions of benefits obtained from the research. The answers were categorized into one of the following four groups: 1) Direct benefit; 2) Indirect benefit; 3) Incentive; 4) No benefit, or benefit has no relation with the research. The direct benefit includes benefits that were caused by the technology itself, and indirect benefit includes benefits that were caused by other factors in the research activities.

Seeds, insecticides and fuels were provided by the researchers to carry out the research activities and are not considered a benefit to farmers in the context of the participatory approach. The benefits were also categorized into six groups: 1) Income and yield increased /livelihood improved; 2) Technical knowledge gained; 3) Inputs provided; 4) Labour saving; 5) Opportunity to access the technology; 6) No benefit. The frequency of replies in each category was tallied.

Results

Changes in yield and cultivated land area of targeted crop

At least one or more farmers responded that they experienced an increase in yield and/or land area across all the research topics. However, in only five research topics were there clearly recognized relationships between the research activity and the reason for the increase. The main reason given for the increase in yield/cultivated land area was the farming techniques learned during the research process, and were not directly related to the specific research topics themselves (Table 9.3). The reason attributed for the increase matched the 'farmer's benefit obtained from the research' for 22.8% of respondents.

Table 9.3 Examples of reasons for increased yield and land area for cultivation, matched with farmer's perceived benefits

Technology developed	Reason for yield/cultivated land area increase, matched with farmer's perceived benefits
Seed treatment with different rate of fungicides on groundnut	• Better weeding and storage practice • Improved soil conservation method (rotation) • Improved weeding method and earthing
New rice variety	• Appropriate frequency of fertilizer input • Row planting for ease of fertilizer application and weeding • Appropriate amount of fertilizer • Seed of improved variety • Improved agronomic practices (weeding and ploughing) • Appropriate frequency of ploughing
New teff variety and row sowing	• Improved management practices (fertilizer, early ploughing, compaction)
Wheat seed purification	• Row sowing, weed control, reduction of seeding rate • Improved irrigation management • Using second-generation purified seed • Row planting and seed purification
Lower seeding rates on rice	• Proper weeding frequency and increased irrigation frequency • Rice knowledge introduced

Farmers' adoption of technology

The proportion of respondents who adopted an introduced technology comprised 34% of the total. The main reason given for an adoption was 'superior crop performance (excluding yield)'. The frequency was much higher than for 'higher yield' (Figure 9.1). The main reason given for non-adoption was 'not remembering the technology', followed by 'the technology is not suitable for farming environment/condition', 'there is a suitable/better alternative', and 'The technology requires more labour and time' (Figure 9.2).

Challenges in cultivation

Challenges encountered in the cultivation of the target crops varied among the farmers' responses, with problems relating to labour, diseases and pests accounting for the majority of reasons given (Figure 9.3). The challenges which were covered in the research accounted for 32% of the total.

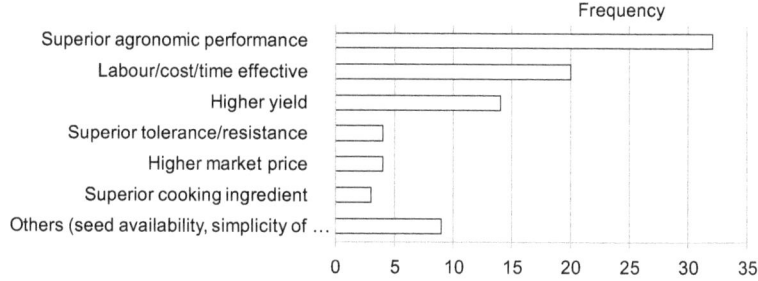

Figure 9.1 Farmer's reasons for technology adoption

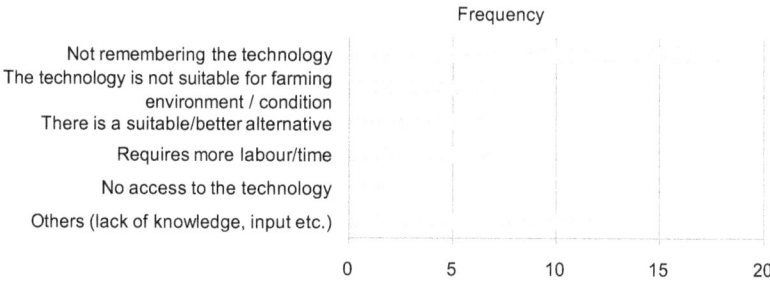

Figure 9.2 Farmers' reasons for non-adoption of technology

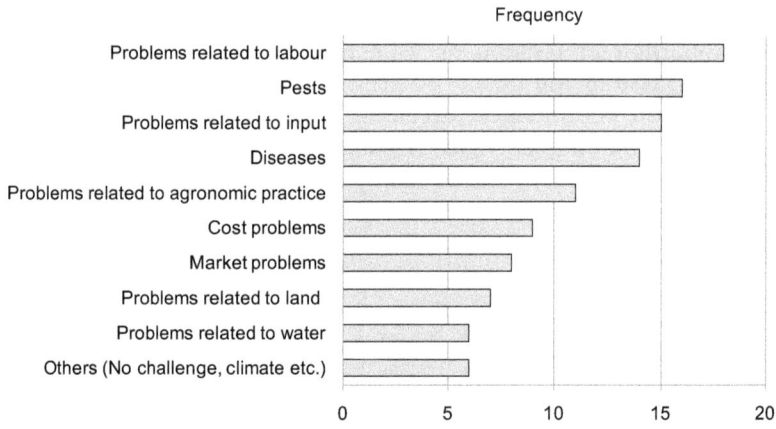

Figure 9.3 Challenges encountered in cultivation of target crop, as perceived by farmers

Farmers' perceptions of the benefits obtained from the research

The technological knowledge gained through participating in research activities was perceived by the respondents to be a major benefit (Figure 9.4). Incentives such as access to the seeds of crop, insecticide and fuel provided by the researchers came second as benefits perceived by farmers (Figure 9.5). The most frequent type of the benefits was indirect followed by direct (Figure. 9.4).

Discussion

Changes in yield and cultivated land area

About a quarter (23%) of farmers interviewed answered that they had achieved increased yields and/or greater land area under cultivation because of the research. As shown in Table 9.3, simple farming techniques which can be adopted by farmers within their existing resources were recognized as contributing to the increases, not necessarily the specific technologies researched. This may be partially explained by either or both of the reasons given for non-adoption: 'The technology is not suitable for farming environment/condition', and 'no access to the technology'.

Adoption of the technology

The FRG farmers had higher technology adoption rates than the non-FRG farmers although the sample number was not large enough for a thorough comparison. Further detailed surveys will be necessary to evaluate whether

FARMERS' PERCEIVED BENEFITS OF FRG-BASED RESEARCH

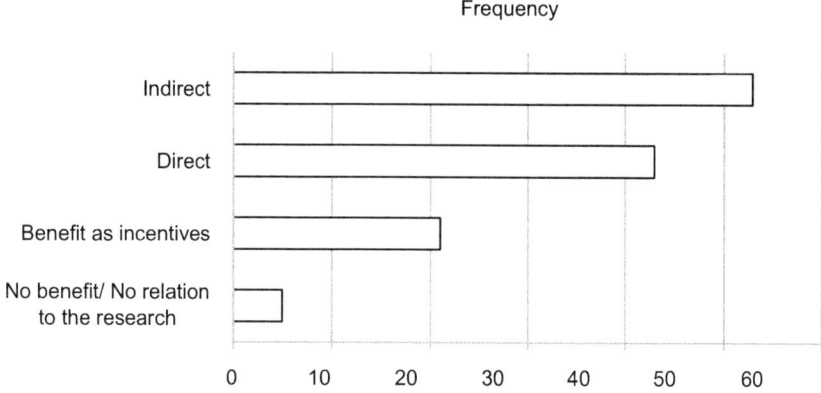

Figure 9.4 Type of benefits obtained from the research

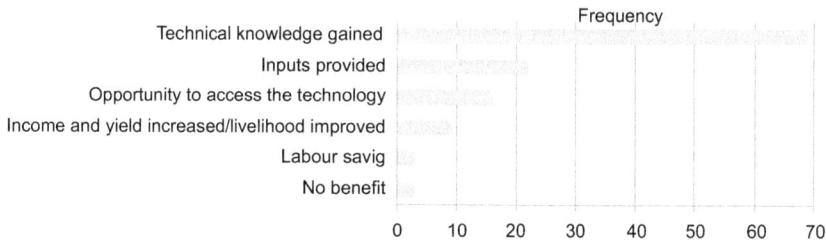

Figure 9.5 Benefits perceived by farmers

technology adoption by 34% of the respondents has any implication for FRG approaches.

The 'superior agronomic performance (excluding yield)' was the main reason given for technology adoption, and its frequency was much higher than 'higher yield'. Expressions given by respondents which were included in the 'superior agronomic performance' category were uniform emergence, higher/lower plant height, larger seed size, higher tillering capacity, heavy grain weight, larger grain size, superior germination performance, faster growing, dominates weed, stronger stalk, big spike. Therefore, higher yield is not always the only criterion for adoption. The farmers evaluate the technology from various aspects may be better of plant morphology and physiology.

The main reason for non-adoption was 'not remembering the technology'. There could be many causes for this. However, when viewed from the perspective of the participatory approach, it might be suggested that communication between farmers and researchers was not carried out well. In the FRG guidelines, the importance of communication between researchers and farmers is highlighted at the beginning of FRG research, for functional linkage and better interaction (Bedru et al., 2009). Through these communications, farmers' impressions of the technology must be reinforced. The role of communication in committing things to memory is scientifically proven (Takatori 1980; Hollingshead 1998). The farmers' perception of 'not remembering the technology' indicates that there is still a shortage of communication skills among researchers and farmers, which should be integrated into further improvements of guidelines and training activities.

Challenges in cultivation of target crop

Farmers' perceptions varied about the challenges encountered when cultivating the subject crop in question. The challenges addressed by the FRG researchers comprised 32% of the total challenges mentioned by the respondents. From the aspect of the FRG approach, this could imply the initial steps of 'identification and prioritization of farmers' problem' and 'research topic identification' need to be more carefully aligned by researchers. This implication seems to be supported by the main reason given for non-adoption of a technology, that 'the technology is not suitable for farming environment/condition'. In the impact evaluation report on the first phase of the project, the weakness 'technology that does not meet farmers' need' was raised (Bezabih, 2009). The inappropriateness of the newly introduced technologies for poor farmers is also reported in several other countries (Hall and Nahdy 1999; Waters-Bayer et al., 2004). Recognizing such mismatches between farmers' perceived challenges and the research topics identified by researchers who have implemented FRG approach activities would certainly be the first step in improving the quality of the research process in future activities, by making greater efforts to meet farmers' actual needs and fully understand their circumstances.

Benefits obtained through the research

Even though a relatively small number of respondents adopted the technology, many respondents answered that they obtained benefits from participating in the research process. The greatest benefit recognized by respondents was that they had gained technical knowledge. The opportunity to access technology was also named as one of the major benefits. Through this it can be implied that FRG-based research contributed to the technical options available to farmers. However, it is difficult to judge if the participatory concepts of the FRG approach influenced these benefits, since the extension activities themselves can also influence farmers. Further investigation to reveal the degree of farmers' participation in research is needed.

Conclusion

A majority of the interviewed farmers perceived that they had obtained benefits from the research process of FRG activities, and some of the farmers achieved increased yields. As against the technologies tested and/or developed through the research process, other simple farming techniques learned in the research process that were compatible with their existing practices and resources were recognized as benefits. In other words, the farmers evaluated and adopted useful technologies based on various aspects relevant or important to them, rather than on just one criterion of production performance. On the other hand, there were cases where participating farmers did not remember the researched technology itself. This may be because the results achieved through using the unadopted technologies did not directly address the challenges faced by farmers or their natural and socio-economic environment, and/ or shortcomings in researchers' skills in communication and facilitation. Important issues for the institutionalization and sustainability of the FRG approach will be further improvements aimed at capacity development, related to matching farmers' felt needs and research subjects, based on the experiences of FRG II, in addition to carefully designing research that takes into account multiple criteria, which might include technical, economic and social aspects, as well as farming system compatibility. In order to confirm the impact of the FRG approach, more precise and thorough socio-economic impact evaluation needs to be conducted, after the completion of an intervention.

About the author

Shingo Takeda was engaged by the FRG II project as an expert in coordination and impact evaluation between 2011 and 2015. He holds a BSc degree in Veterinary Medicine from Azabu University, Kanagawa, and an ME in Bioengineering from Soka University, Tokyo. He has lived and worked in Thailand, Djibouti, the Republic of Palau and Ethiopia.

References

Bedru, B., Berhanu, S., Endeshaw, H., I. Matsumoto, M. Niioka, K. Shiratori, Teha, M. and Wole, K. (2009) *Guideline to Participatory Agricultural Research through Farmer Research Group (FRG) for Agricultural Researchers*, FRG II, EIAR/OARI/JICA, Addis Ababa.

Bezabih, E., (2009) *Evaluation of Impact of Farmers Research Group Activities in the Rift Valley of Ethiopia*. Project document of Farmer Research Group Project, EIAR, OARI and JICA, Addis Abana.

Catley, A., Burns, J., Abebe, D., and Suji, O. (2007) *Participatory Impact Assessment: a Guide for Practitioners*, Tufts University, Feinstein International Center USA.

Hall, A. and Nahdy, S. (1999) *New methods and old Institutions: The "systems context" of farmer participatory research in national agriculture research systems. The case of Uganda.* Network Paper Number 93, Agricultural Research and Extension Network London, UK.

Hollingshead, A.B. (1998) 'Communication, Learning, and Retrieval in Transactive Memory Systems', *Journal of Experimental Social Psychology*, 34 (5): 423-442.

Jones, A.L. (2001) 'Assessing the Impact of Participatory Research and Gender Analysis', in J.A. Ashby, & L. Sperling (eds.), CGIAR Program for Participatory Research and Gender Analysis Cali, Colombia.

Sato, I. (2008) *Qualitative data analysis method*, Shitsuteki Data Bunsekihou (Japanese), Shin'yosha, Tokyo.

Seo, T. (2016) 'Farmer Research Group implementation processes: guidelines, training and research', in Alemu, D., Nishikawa, Y., Shiratori, K. and Seo, T. (eds), *Farmer Research Groups Institutionalizing Participatory Agricultural Research in Ethiopia*, pp. 25–42, Rugby, UK: Practical Action Publishing <http://dx.doi.org/10.3362/9781780449005.003>.

Takatori, K. (1980) 'The role of communication in the memory process', *Japan Journal of Educational Psychology*, 28 (2): 108-113.

Waters-Bayer, A., Van Veldhuizen, L., Wettasinha, C. and Wongtschowski, M. (2004) 'Developing partnerships to promote local innovation', *The Journal of Agricultural Education and Extension*, 10: 143-150.

PART III
Institutionalizing Farmer Research Groups in Ethiopia

CHAPTER 10

The participatory approach and FRG: the institutionalization process within the Ethiopian agricultural research system

Dawit Alemu and Kiyoshi Shiratori

Abstract

This chapter presents the historical evolution of participatory research in the Ethiopian agricultural research system (NARS) and its institutionalization process. Different participatory approaches have been implemented since the early 1970s. The first attempt was Farming System Research (FSR) followed by different crop improvement programmes that embedded participatory approaches, such as Participatory Plant Breeding (PPB) and Participatory Variety Selection (PVS). There was also a parallel endeavour in the second half of the 1990s to strengthen research-extension linkage, which promoted Farmer Research Groups (FRGs). The institutionalization of the FRG approach took on different dimensions: (i) the establishment of research-supported FRGs, (ii) the development of FRG approach guidelines and associated capacity building for researchers, (iii) the assignment of FRG training hubs, and (iv) the development of a training curriculum on the FRG approach within universities. The challenges faced are sustaining capacity once developed, ensuring adequate finance, and the non-uniformity of the FRG approach between the different research centres.

Keywords: Ethiopian agricultural research system, participatory research approach, Farmer Research Groups, institutionalization, research and extension

Introduction

The need to promote participatory research in the Ethiopian agricultural research system has been recognized since the early 1970s. Since then different types of participatory approaches have evolved and been implemented. The first approach implemented was Farming System Research (FSR), which has played an important role in promoting different types of on-field research activities by engaging farmers and other stakeholders especially in technology package evaluation. In the second half of the 1990s, the Farmer Research Group (FRG) approach was adopted by members of the national agricultural research system (NARS), mainly by the Ethiopian Institute of Agricultural Research (EIAR) and the regional agricultural research institutes (RARIs), with the objective of making agricultural research activities responsive to the needs of the relevant farmers by ensuring that judgements made about the appropriateness of any introduced technologies were made by smallholder farmers based on their particular socioeconomic circumstances (FDRE, 1999).

The process of institutionalizing the participatory approach can take on different dimensions. It can follow an officially recognized process through introducing regulation, or it can follow an informal process, where it becomes accepted by mainstream researchers as the favoured technical approach when conducting research. Following the demonstrable achievements from applying the participatory approach within the research system, there have been different efforts to embed the different approaches within the NARS. During this process, the different approaches have followed different pathways. This paper looks at how the institutionalization process of the different participatory approaches has evolved and been incorporated into a well-established system with due emphasis to the FRG approach in EIAR and RARIs. The chapter first presents the historical trends in institutionalizing the participatory approach in the research system, followed by a discussion regarding how the FRG approach has been institutionalized. The conclusion considers the key challenges and lessons learned during the institutionalization process.

Historical trends

Different participative approaches to agricultural research have been implemented following recognition of the need for them in the 1970s (Aberra and Fasil, 2005). Under Ethiopian conditions, the first attempt at promoting participatory research was FSR, which recognized the low success rate of conventional research approaches and also the fact that agricultural production takes place within complex farming systems that require more holistic research approaches to be more relevant to farming communities. Different crop improvement programmes that embedded participatory approaches, such as Participatory Plant Breeding (PPB) and Participatory Variety Selection (PVS), were also implemented around that time. The main target of the PPB and PVS approaches was to promote adoption of new varieties through: 1) increasing farmers' awareness of different crop variety traits; 2) building farmers' capacity in evaluating crop varieties for different traits and incorporating farmers' opinions in the variety selection process; 3) improving access to different varieties for farmers' own experimentation and selection; and 4) creating business opportunities in seed marketing for small-scale farmers using well-accepted varieties (Alemu, 2011).

The FSR institutionalization process

In 1976, the then Institute of Agricultural Research, now the Ethiopian Institute of Agricultural Research (EIAR), started to promote participatory research through on-farm research activities through its Department of Agricultural Economics (DAE). The aim was to identify farmers' circumstances and formulate a package of technologies for on-farm testing, and to demonstrate the available technologies in farm communities (Kean and Ndiyoi, 2000). This was then followed by a formal arrangement regarding the implementation of FSR by assigning researchers with different disciplines and renaming the department the Department of Agricultural Economics and Farming System Research (DAEFSR). This arrangement existed between 1984 and 1988 and only two agronomists worked with economists under the DAEFSR aegis. It was recognized that once researchers from different disciplines were exposed to the FSR approach, it was possible to initiate inter-departmental research projects. It was no longer necessary to base a multidisciplinary FSR team within DAEFSR. The agronomists were transferred to the Department of Agronomy and the DAEFSR reverted to DAE, with FSR as a function of the institute as a whole, rather than the responsibility of a single department (Kean and Ndiyoi, 2000). Since the 1990s, the implementation of FSR activities through multidisciplinary teams per se has not persisted. However, on-farm research and some of the key components of FSR-like diagnostic surveying have continued. The key mechanisms for ensuring a multidisciplinary approach in on-farm research activities have been FSR training programmes, joint reviews of research proposals, annual planning meetings, and a formal directive outlining the mode of collaboration. The coordination and institutional arrangements required to ensure the effective engagement of researchers of different disciplines in on-farm research activities remains a critical challenge.

The PPB and PVS institutionalization process

The crop breeding and genetics research in EIAR focused in the early years on developing varieties responsive to high external inputs. In the 1990s, the research focus moved towards a participatory and multidisciplinary approach, with the major emphasis placed on on-farm research with the full participation of farmers (Bayeh and Berhane, 2011). In this regard, PVS and PPB within EIAR were recognized as applications of farmers' participatory research (FPR) in variety development and dissemination activities. Their application within the system was a response to criticism of formal research methods for their failure to produce technologies appropriate for resource-poor farmers, and the centralized, institutional approach they adopted (Getachew et al., 2008; Witcombe et al., 2005). The application of FPR quickly demonstrated successes. For instance, the teff breeding programme has released the popular teff variety 'Quncho' and the beans breeding programme has contributed to the popularization of preferred bean varieties through PPB and PVS (Getachew et al., 2008; Teshale et al., 2005).

PVS and PPB follow procedures where both researchers and farmers are involved in the process. The steps in the process are: 1) researchers' selection of promising lines following a formal breeding procedure, 2) farmers' on-station selection from the promising lines, 3) researcher performance evaluation under national variety trials of varieties selected by farmers, (iv) identification of candidate varieties for National Variety Release Committee (NVRC) evaluation and for on-farm trials by both farmers and researchers, (v) on-farm trials and farmers' preference ranking, and (vi) formal evaluation and release by the NVRC (Getachew et al., 2008). Therefore, the institutionalization process of PVS and PPB in EIAR is within the overall shift of crop improvement programmes towards a participatory and multidisciplinary approach where PVS and PPB have become the key procedures.

The institutionalization of the FRG approach

In the international arena of promoting participatory research, the research group approach has rapidly gained ground and attracted the attention of many research and development organizations to address agricultural and natural resource management problems. Notable examples of group-based participatory research approaches that have spread widely include the local agricultural research committees in Latin America, farmers' field schools in Asia and Africa, and FRGs in southern and eastern Africa (Sanginga et al., 2006).

In Ethiopia, along with the promotion of participatory research, there has been a parallel endeavour to strengthen the research-extension linkage. This endeavour was instrumental in further strengthening the promotion of participatory research. In the second half of the 1990s, the FRG approach was adopted by the NARS, mainly by EIAR and RARIs as an innovative strategy to make agricultural research and extension activities responsive and relevant (FDRE, 1999). An FRG is a group that farmers form voluntarily to undertake experiments on their own fields. The groups are formed around farmers' production constraints as identified and prioritized by farmers themselves. An FRG may have a chairperson and secretary elected by members (the only proviso being that there must be a women's representative), and a membership which comprises those people who register with the group for a particular season's activity. The membership of FRGs is not fixed. People flow in and out of them, although a core of members provides continuity from one season to the next. The groups, however, have a collective memory, which individuals do not have. The essence of FRGs is to make agricultural research and extension client-oriented and thereby develop informal, collaborative relationships and partnerships which will in the final analysis enhance the impact of research and extension activities (Teklu, 2007).

In general, the institutionalization of the FRG approach has followed two routes. The first is the formal consideration of the approach in the research system along with the establishment of FRG training hubs. The second route is the scientific recognition of the importance of the approach within the research system, initiating the application of the approach by researchers themselves, without the formal directives. Specifically, the wider recognition of the approach by researchers with a social science research background at each research centre of EIAR and the RARIs plays a crucial role in the institutionalization of the approach within the NARS.

Within the NARS, six FRG training hubs have been established. These are the Melkassa and Werer agricultural research centres of the EIAR, Adami Tulu Agricultural Research Centre of Oromiya Agricultural Research Institute and Mekele, Hawassa and Bahir Dar universities. Interventions with these have been: (i) the provision of training of trainers, undertaken by the FRG II project to researchers and instructors with different specializations and research experience, (ii) the provision of required information in the form of different publications including manuals and relevant research reports, (iii) support for the establishment of required training facilities, and (iv) official communication to respective managing bodies about the establishment of the hubs, outlining responsibilities.

Research and extension divisions at each research centre of EIAR and RARIs used to take the initiative to set up and facilitate the establishment of FRGs. The interaction between farmers, researchers and extension workers revolves around joint planning, joint experimentations, joint field visits, field days, and study visits, all of which are based on the basic principles of participatory approaches. Following the Business Process Re-engineering process, the Agricultural Economics, Extension and Gender Research Directorates (AEEG) at each federal and regional research centre[1] are now responsible for initiating and promoting the FRG approach. In some RARIs where the AEEG research programmes are weak or have been abolished, the promotion of the FRG approach is undertaken in an ad hoc manner.

As noted earlier, the rationale behind the formation of FRGs is to make agricultural research and extension client-oriented and thereby develop informal, collaborative relationships and partnership, which will enhance the impact of research and research-extension[2] activities. According to FDRE (1999), carrying out research and research-extension through FRGs helps to:

- exert pressure on research and extension for the development of demand driven technology by influencing research and extension interventions;
- optimize researchers' and extension workers' time and effort required to interact with farmers;
- facilitate group actions such as natural resource conservation and management and provide a vehicle for researchers and groups of farmers to work together;
- re-orient the research agenda and operational culture of research and extension towards farmers' priority needs, to share responsibilities with research and extension in problem identification, testing and transfer of technologies, and to encourage women farmers to participate in technology generation, evaluation and dissemination;
- ensure a sustainable, informal technology diffusion network between the groups in adjacent villages and among the farming community.

The Farmer Research and Extension Group (FREG) approach, promoted and widely implemented by the Regional Capacity Building Project, emphasized the need to communicate the importance of extension in technology evaluation and adaptation. The emergence of the term FREG caused many researchers and practitioners to raise questions regarding the differences between the FREG and FRG approaches. Since the FREG adopted the FRG concept and was implemented mainly by extension workers, FREG focuses more on the dissemination of improved technologies. Although the FRG approach aims for its outputs to be disseminated by extension work, its focus is specifically on generating new technical information in which process researchers have important roles. The participation of farmers and DAs in FRG-based research effectively combines their expert understanding of agro-biology and the socioeconomic nature of the area where a technology is being trialled.

Challenges and lessons in institutionalizing the FRG approach in NARS

Though the institutionalization of the FRG approach has progressed well both formally and informally, there are challenges to meet to ensure sustainability. These are: (i) sustaining the developed human capacity, hampered by staff turnover, (ii) ensuring adequate finance to undertake planned training and on reviewing the evolution of the approach, and (iii) the non-uniformity of approach taken to FRG research among the different research centres, especially the balance of consideration given to the approach as a research tool, extension tool, or both research and extension tool.

As indicated above, one of the strategies followed in institutionalizing the FRG approach was the establishment of training hubs that consider the diverse actors involved in the NARS. In total, six training hubs were established, three from the research institutes and three from universities. The key challenge in these hubs has been coping with staff turnover, resulting in a loss of knowledge about the approach, experience of running adult training programmes, and the capacity to mobilize resources. From the six training hubs, only four, namely Melkassa and Adami Tulu agricultural research centres and Mekelle and Hawassa universities, were able to run complete training programmes as per their annual plans.

Staff turnover also has implications for advancing the approach, using emerging lessons and experiences gained as FRG-based research activities continue. In addition, following the adoption of group action by the Government of Ethiopia for rural development in general and agricultural extension in particular, there is huge interest by extension officers to adapt and apply FRG approach principles in their extension efforts. This has created a debate about whether the FRG approach is part of research or extension, resulting in turn in a wider tendency of researchers to use FRGs as if they are purely an agricultural extension approach, limiting the research dimensions that can be explored by both researchers and participating farmers. It has further resulted in a lack of attention being paid to scientific procedure, a lack of scientific evidence, and intrusive promotion of modern technologies, which are often inappropriate.

The key lessons learned in the implementation of the FRG approach and its institutionalization indicate that it is important to further strengthen the established hubs, continue with planned capacity building to researchers on the FRG approach, develop and improve the interface between FRG and FREG, and continue to incorporate the approach systematically within national agricultural research strategies.

Conclusion

This chapter has presented the institutionalization process of the FRG approach based on the experiences of the FRG II project. Following the wider acceptance of participatory research within the NARS, different approaches have been implemented. In the second half of the 1990s, strengthening the promotion of participatory research in the form of FRGs was adapted as one of the mechanisms to enhance the research-extension linkage. The FRG approach was an innovative strategy that has been adopted by the NARS to make agricultural research and extension activities responsive and relevant to farmers, prioritizing the involvement of smallholder farmers in an organized manner in the selection of research and extension priorities, research planning, and implementation.

The institutionalization of the FRG approach followed two dimensions, where the first was the formal institutionalization of the approach through the establishment of FRG-approach training hubs. The second was the scientific recognition of the importance of the approach by researchers themselves without formal directives. Wider recognition has resulted in consideration given to the FRG approach in research activities, mainly by researchers with social science backgrounds within the NARS.

There are challenges in further strengthening the application of the approach, in terms of the considerable turnover of trained and experienced researchers and a shortage of financial resources to undertake planned FRG-based activities. However, the institutionalization process, mainly via the establishment of training hubs, and the availability of FRG approach information (manuals and related documents) are expected to ensure the continued application and consideration of the approach within the NARS.

Notes

1. Some RARIs have different naming but the roles and responsibilities are very similar.
2. Research-extension activities are performed by the NARS to create a demand for generated technologies and wider awareness of the mainstream extension system.

About the author

Dawit Alemu is Director of the Agricultural Economics, Extension and Gender Research Directorate of the Ethiopian Institute for Agricultural Research. He has been associated with EIAR since 1999 as a Senior Researcher and Coordinator. His research focuses on agricultural marketing with an emphasis on agricultural inputs.

Kiyoshi Shiratori is a consultant specializing in rural development at Kaihastu Management Consulting. He is a visiting professor at the Graduate School of Asian and Africa Area Studies, Kyoto University. He was a Chief Advisor in the FRG I and FRG II projects in Ethiopia.

References

Bayeh M. and Berhane L. (2011) 'Barley research and development in Ethiopia – an overview', in: Mulatu, B. and Grando, S. (eds.), *Barley Research and Development in Ethiopia*, Proceedings of the 2nd National Barley Research and Development Review Workshop. 28-30 November 2006, HARC, Holetta, Ethiopia. ICARDA, PO Box 5466, Aleppo, Syria. pp xiv + 391.

Getachew B., Hailu T., Anteneh G., Kebebew A. and Gizaw M. (2008) 'Highly client-oriented breeding with farmer participation in the Ethiopian cereal tef [*Eragrostis tef (Zucc.) Trotter*', *African Journal of Agricultural Research* 3 (1), pp. 22-28.

Kean, S. A. and Creasy, N. (2000) 'The institutionalization of FSR in East and Southern Africa: An overview', in: M. Collinson (ed.), *A History of Farming Systems Research*, pp. 191-200, Rome.

Martin, A. and Sherington, J. (1997) 'Participatory research methods – implementation, effectiveness and institutional context', *Agricultural Systems*, 55, pp. 165-216.

Teshale A., Girma A., Chemeda F., Bulti T. and Abdel-Rahman M. (2005) 'Participatory Bean Breeding with Women and Small Holder Farmers in Eastern Ethiopia', *World Journal of Agricultural Sciences*, 1 (1), pp. 28-35.

Witcombe, J., Petre, R., Jones, S. and Joshi, A. (1999) 'Farmer participatory crop improvement. The spread and impact of a rice variety identified by participatory varietal selection', *Experimental Agriculture*, 35, pp. 471-487.

CHAPTER 11
The challenges of FRG-based research: attitudes, capacity and institutional arrangements

Dawit Alemu, Taku Seo, Terutaka Niide, Shingo Takeda, Kiyoshi Shiratori and Yoshiaki Nishikawa,

Abstract

This chapter presents the key challenges faced in promoting Farmer Research Group (FRG)-based research activities. Supported research activities were identified based on a call for proposals and implemented with technical support from the FRG II project staff. The experience gained shows that the success and failures of FRG-based research activities have been highly influenced by: (i) the individual researchers' attitudes towards farmers and their own research activities, (ii) the researchers' technical capacity, (iii) the institutional culture where the researchers are based, and (iv) in some cases, the nature of the FRG research activities themselves and unexpected events such as drought. These challenges suggest considering measures to improve researchers' technical capacity in aligning conventional with participatory approaches, and changing attitudes and institutional arrangements within the respective institutes to better promote FRG research.

Keywords: Ethiopia, Farmer Research Groups, researchers' attitude, researchers' capacity, institutionalization

Introduction

Farmer Research Group (FRG) research activities have been implemented by diverse researchers from different research institutes including the Ethiopian Institute of Agricultural Research (EIAR) research centres, regional research centres and universities with agricultural faculties. In order to ensure wider coverage of different stakeholders, research ideas and geographic areas, the FRG-based research activities supported by the Project for Enhancing the Development and Dissemination of Agricultural Innovations through Farmer Research Groups (FRG II) were identified based on a 'call for proposals' within identified thematic areas. Once feasible research proposals were selected by a panel of senior researchers, the researchers were invited for proposal review meetings. These meetings were often attended by senior researchers, and proposals revised in terms of their main content, methodology, and targeted sites, based on discussion and common understanding.

In the implementation process, technical backstopping and monitoring and evaluation were performed to ensure consistent high quality. Following the completion of the research activities, final research output review meetings were organized to share experiences and upgrade the scientific content of the final reports. The revised research reports are formally published for wider circulation once approved by reviewers. In some cases, extension materials have been developed from the research results.

Any agricultural research, either conventional or participatory, has its challenges and opportunities. FRG-based research activities deal with farmers' priority issues through the involvement of multidisciplinary teams of researchers, farmers, and development agents (DAs, frontline extension workers who work directly with farmers). The implementation challenges of FRG-based research activities emerge from the level and intensity of collaboration of these actors and from individual actors themselves. In addition, the mechanisms and procedures followed in the design of the research play an important role for achieving the research output. This chapter presents the main challenges and opportunities faced in the process of implementing FRG-based research activities based on the FRG II experience over a project period of five years.

Major challenges in implementing FRG-based research activities

The challenges encountered during the FRG approach implementation process have been factors in the failure of some of the FRG research activities funded by the project. The main issues have been: (i) researchers' technical capacity, (ii) attitudes and research ethics, (iii) organizational culture and working practices, and, in some cases, (iv) the nature of the FRG-based research activities themselves. These challenges are discussed in detail below.

Researchers' technical capacity

The FRG-based research approach considers two dimensions of research. The first is the application of scientific research methods, principles, associated data collection, and analysis (often recognized as conventional research), and the second is the application of methods to ensure farmers' and other stakeholders' participation in the whole process, beginning with research design, through to the validation of research results, which is often called the participatory approach. Some argue that the participatory approach without associated scientific evidence can also ensure technological innovation, based on the assumption that farmers can select beneficial technologies and ignore others. However, in reality, research outputs without scientific evidence may lead to promoting inappropriate technologies. Unless augmented with scientific evidence, the research result may ignore the important aspects of modifying and fitting technologies into farmers' specific situations.

In the implementation of FRG-based research, it is expected that the researchers will be well-equipped with knowledge of scientific research methods, principles, and associated data collection and analysis skills to accurately document the research output. This is because FRG-based research is about embedding both participatory principles and scientific research methods into the entire research process. Experience in promoting FRG-based research activities revealed three possible scenarios.

The first is what would be normally expected where researchers apply participatory principles and approaches to their conventional research methods from initial design, through implementation to validation of the research output (see Box 11.1).

The second scenario is where researchers apply inappropriate scientific research methods in combination with the participatory approach based on FRG principles (see Box 11.2). In this case, the information generated often meets challenges to accurately demonstrate the scientific evidence. These situations have not been documented well in previous studies of participatory research.

The third scenario is where researchers completely ignore the conventional methods and try to base the research solely on the participatory approach without applying the key principles of the FRG approach. This often leads to

---

Box 11.1 Successful implementation of FRG-based research: the case of determination of fertilizer rate for upland rice in Benishangul Gumuz region

In Ethiopia, the blanket fertilizer rate recommendation has been challenged for its inappropriateness given the diversity of soil types and crop requirements. Cognizant of this, a team of researchers from Assosa Agricultural Research Centre of EIAR (an agronomist, an economist and a plant scientist) implemented an FRG-based research project to determine the optimum Nitrogen (N) and Phosphorus (P) fertilizer application rate for upland rice. The experiment started with the formation of an FRG, where members were provided with explanations about the experiment, agreed the topic and were given the necessary training for implementing the experiment. Though the experiment was conducted on member farmers' fields, the design followed strictly scientific methods (factorial experiment laid out in a randomized complete block design with three replications). It also incorporated farmers' suggestions – for example, the application levels for N (0, 46, 92, 138 kg/ha) and for P (0, 23, 46, 69 kg/ha P_2O_5). The research evaluation was made jointly by FRG members, extensionists and researchers on a regular basis. The results indicated that using a 92 kg/ha of N and 46 kg/ha of P_2O_5 fertilizer application rate consistently produced a higher yield and better resistance to major rice diseases. The recommendation was well-accepted by FRG member farmers based on their own observations and analysis of growth and yield. The key reasons for successful implementation of this FRG-based research are: (i) appropriate involvement of FRG member farmers from research design through to evaluation, (ii) application of appropriate scientific research methods, and (iii) the right composition of the research team especially in terms of relevant discipline and senior experience.

> **Box 11.2 Implementation of FRG-based research, where inappropriate scientific methods are applied: the case of participatory management of storage to prevent insect pests of maize**
>
> Reduction of post-harvest losses is considered to be one of the three pillars of ensuring food security, alongside increasing productivity and enhancing production. To address this, a team of researchers implemented FRG-based research on the participatory management of storage to prevent insect pests of maize in Jimma zone, Ethiopia. The team members were an entomologist, an economist and a cereal breeder. The team initially proposed: (i) to identify a list of storage management options from existing literature and from farmers' indigenous knowledge, (ii) to prioritize with farmers the best potential options, and (iii) to design a scientifically sound experiment using the prioritized options, to be applied to all FRG sites as replications. However, during the implementation, distracted by farmers' individual interests, the options (treatments) applied by the research team were different across the sites and no scientific method/research design was applied. This resulted in the selection of the treatments based on individual observation that varied across the research sites. The farmers' involvement in the process of implementation was well managed, but the research result could not be scientifically documented because of the faulty application of scientific method. The key reasons for misapplying scientific method is the limited capacity of researchers coupled with the temptation to fulfil farmers' individual interests in different research sites.
>
> There was also vary fundamental misunderstanding on statistics. The research team explained the process of sample farmer selection as 'farmers were randomly selected', however, when the project asked the research team what they actually did, the explanation was 'farmers were selected using a random number table'. The explanation was followed with: '(1) random numbers were given to the names on the list of farmers, (2) random number on the table was pointed by the extension agents without looking at the numbers, and (3) (2) was repeated to select all samples'. Just like this explanation, researchers often state that samples were selected 'randomly' without following appropriate procedure to satisfy requirements as 'randomization'. Through our experiences, there was only one researcher who could clearly explain the difference between 'unsystematic selection' and 'random selection', and who actually practised the necessary procedure to claim 'random selection'.
>
> Additionally, the differences between the treatments of the harvested maize in the various locations were explained as follows: 'The sacks containing maize were positioned separately, side by side on stones 14 cm (at the first site), on wood 14 cm (at the second site), and on wooden bench 43 cm above the ground (at the third site)'. All of the data from different locations were then combined, ignoring the differences in storage arrangements, despite the research team being aware of it.

more development work, rather than research resulting in a clear demonstration of the scientific evidence and a full explanation of the advantages of farmers' participation (see Box 11.3).

The reasons for these second and third cases are often associated with the researchers' lack of experience in linking scientific research with participatory research in general and with the FRG approach in particular. This challenge necessitates giving attention to building researchers' capabilities in aligning scientific research with participatory through a clear understanding that the two are not alternatives but complementary. The participatory approach builds on scientific research methods to demonstrate scientific evidence.

> **Box 11.3 FRG-based research where no scientific method is applied: the case of the participatory improvement of forage seed production in pastoral and agro-pastoral areas**
>
> In pastoral areas of Ethiopia, livestock plays a key role in the livelihoods of the communities. Two of the main challenges in livestock production are water and feed availability. To help address pastoralists' problem with feed in the Gursum area of Somali region, an FRG-based research project on promoting forage seed production for alfalfa and *panicum antidotale* was implemented by a group of researchers comprising two animal scientists, one extension researcher and one economist, all of them with BSc degrees. The researchers participated in the review meeting for proposal improvement and made suggestions regarding the scientific methods to be applied, and how the participation of pastoralists should ensure identifying critical issues in forage seed production. However, when the research came to be implemented, the researchers treated it purely as development work, where pastoralists/agro-pastoralists were to be trained on the production of forage crop seeds. To demonstrate the project's achievements, a field day was organized for wider promotion of forage seed production. The main reasons for the inappropriate implementation of this case were: (i) the limited participatory research capabilities and exposure to the FRG-based approach of the researchers involved, (ii) the limited commitment to adhere to such research approaches, and (iii) the organizational culture and work patterns of the establishments employing the researchers, which do not promote mentoring by and engagement with senior researchers.
>
> The final report was concluded with symbolic statements such as the following, which reflects the fact that the researchers had completely missed the research component: 'Participatory on-farm research requires the active participation of major stakeholders particularly DAs and farmers. Though it is only beginning, it was made possible to increase farmers' level of understanding, participation and knowledge about production of improved forage seed (alfalfa and *panicum antidotale*). This has increased their interest and confidence to engage in improved forage seed production. The knowledge and skill of site DAs have also improved. Different stakeholders such as non-FRG farmer members were made aware of the possibility of producing improved forage seed production in the local area.' None of the statements above was supported by valid evidence, and there was no observation of the collected data, either from the trial field or participating pastoralists.

Researchers' attitudes towards research ethics

Before presenting the observed challenges in the area of research ethics, it is important to frame the dimensions of research ethics. 'Research ethics' refers to the expected conduct of researchers to ensure the claimed research result is valid as sound, scientifically based knowledge. To be considered as such, a research output needs to extend human knowledge beyond what is already known. This is achieved only if the knowledge generated through the research process enters the domain of science after it is presented to others in such a way that they can independently judge its validity (peer review) (NAS, 2009). This is commonly carried out through presentations to facilitate exchange of views, followed by publication in some sort of reputable publication – usually a peer-reviewed journal. A failure to work with established research ethics is closely associated with research misconduct, which is in turn linked with so-called FFP (fabrication, falsification, or plagiarism) in proposing, performing, or reviewing research, or in reporting research results. The experience of FRG-based research indicates limited misconduct in terms of FFP. The issues are more related to some researchers' behaviour in terms of their limited openness to scientific discussion – where one accepts and/or convinces based on the facts presented – and their limited commitment to address emerging research challenges.

In order to promote ethical standards in FRG-based research activities, the FRG II project has put in place a number of measures with an emphasis on promoting rigorous review, monitoring and evaluation of all research activity. The project has been involved in all the research processes, starting with the proposal reviews to the reviews of research reports, along with technical backstopping and monitoring and evaluation activities. In this process, major challenges relating to researchers' attitudes towards research ethics and their level of commitment were observed, specifically for projects that were at some point terminated, which was 47% of the total 43 projects supported. Specifically, the major challenges encountered were related to: 1) limited openness to scientific discussion, 2) limited willingness or ability to consider serious suggestions and comments given during the review and field-level monitoring and evaluation process, 3) limited commitment to adhere to the agreed procedures and requirements, and 4) limited commitment to deliver the final output, which could be all or part of research report-writing, extension material and making available raw data.

In support of strengthening the researchers' ability in scientific argument, FRG II has been creating opportunities for scientific discussion, starting from the proposal review up to the final report review. These have been held in the presence of different research teams experienced in FRG-based research, along with senior researchers from relevant fields of specialization. In these forums, it was observed that many researchers were not only unwilling to engage in scientific discussion, but there was a tendency to personalize arguments, and unwillingness to share data among FRG team members. These issues are associated with limitations in: (i) the capacity to manage data in a suitable form for sharing, (ii) the essential willingness to share, and (iii) lack of recognition of the importance of keeping and sharing data.

The review forums were organized to create opportunities for researchers to share experiences, collect new ideas and listen to suggestions for improving the quality of their respective research activities. The experience of the FRG II project, however, indicates that there were researchers who were not willing seriously to consider any new ideas or suggestions about revising either research methods or approach. In this regard, FRG II had to create an administrative 'enforcement mechanism', such as the withholding of research funds, not in itself desirable or ethical. Even with such measures, there has been a common 'lip-service' approach, without any real adjustments on the ground in the implementation of research activities.

Implementation of any research activities has to follow certain procedures and fulfil requirements based on the type of research, the source of funding and institutional arrangements. The experiences of FRG II have indicated that limited commitment among researchers coupled with inconsistencies in organizational implementation made it difficult, if not impossible, to adhere to the agreed procedures and requirements.

Organizational culture and working practices

Organizational culture refers to a series of attitudes and behaviours adopted by all employees of a certain organization, which affect its functioning and total well-being and is often considered as the 'glue' that holds an organization together and the 'compass' that provides guidance and direction (Dimitrios and Koustelios, 2014; Tharp, 2009).

Inappropriate, or a lack of, clear work modalities and organizational culture play an important role in and seriously contribute to the failure of research activities. The experience of FRG-based research activities shows that the lack of commitment to an agreement about research team operation, responsibility sharing and institutional accountability have been key factors in the failure of a number of research activities that ended without any output.

Given the high research staff turnover in almost all research institutes, including universities, a number of FRG-based research activities have ended without finalizing a research report or clear output. This is partly associated with the limited commitment of researchers and lack of a system in the respective institutes to ensure the delivery of the final output before researchers leave. There also seemed to be serious lack of concern about handover whereby a number of researchers who obtained an opportunity for long-term study or other employment left the institute without the proper transfer of the research activities and associated resources, including research data, to other research team members. Such a lack is particularly problematic with handling farmers' indigenous knowledge, which requires careful communication and mutual trust. The experience of FRG II indicates that 16 out of 43 research projects had a transfer of principal researchers, which contributed to the late delivery of research outputs or termination of the entire activity. This implies the need to strengthen the different institutes' research management procedures, along with the need to give due attention to the composition of research teams from the outset to include senior researchers.

Nature of the FRG research activities

FRG-based research activities deal with farmers' priority issues through the involvement of a multidisciplinary research team, a group of farmers and DAs. Furthermore, the research activities are often undertaken in an unpredictable environment. Researchers are therefore expected to have the capacity to manage different group interests and be able to address in a timely manner any emerging challenges or external constraints such as drought, pests, and even wild animal attacks. This often requires long-term experience and previous exposure to these challenges, which may not be the case with younger researchers.

This strongly suggests the need for engaging senior researchers in all research teams, to ensure the smooth transfer of skills when addressing any 'unexpected' research challenges. This will also have the advantage of exploiting the potential for mentoring, and the delivery of quality research output(s).

Conclusion

Agricultural research, either conventional or participatory, has its own challenges and opportunities. Participatory research in general and the FRG approach in particular face challenges that emerge from the nature of the research itself. The process involves the identification of farmers' priority problems, the design of appropriate research methodologies, considering both scientific and participatory methods, and managing the interests of the different groups. This all has to be done within the framework of ensuring farmers' continued contributions throughout the experiment and their involvement in the improvement of the research.

In order to ensure proper design, implementation and addressing any challenges, the FRG II project put in place procedures for all FRG-based research activity. These included: prior identification of thematic areas; a call for proposals based on the identified thematic areas; evaluation and selection of feasible proposals; upgrading the contents and methods of selected proposals through review; provision of technical backstopping in the process of implementation; and review of the research results/reports before wider dissemination. These procedures were followed and implemented using criteria that ensure the application of FRG approach principles. However, there were a number of challenges faced that have contributed to the failure of a number of FRG research activities.

In order to address the major challenges observed with regard to research attitude, capacity and institutional arrangement in the process of implementing an FRG project, the following key measures are suggested:

- There is need to give attention in building researchers' capacity in aligning and consideration of conventional/scientific research methods with participatory research principles like the FRG approach as the two are not alternatives but complementary methods to demonstrate scientific evidence in a demand driven manner.
- Research ethics need to be strengthened through careful follow-up of research procedures to ensure the openness of scientific discussions.
- Organizational culture and working methods need to guide proper research team operation, responsibility sharing, and institutional accountability.
- Owing to the nature of the FRG research activities, engagement of senior researchers and mentoring of young researchers is highly beneficial

to nurture the skills required for coping with the challenges associated with on-farm and multi-actor research. This is a key issue that must be inculcated in the culture of all research institutes.
- Challenges can also be considered as opportunities if there is a clear understanding of them, and precautions can be taken starting right from proposal selection, through research team composition, to addressing the possible research capacity gaps.

About the authors

Dawit Alemu is Director of the Agricultural Economics, Extension and Gender Research Directorate of the Ethiopian Institute for Agricultural Research. He has been associated with EIAR since 1999 as a Senior Researcher and Coordinator. His research focuses on agricultural marketing with an emphasis on agricultural inputs.

Taku Seo works with the Smallholder Horticulture Empowerment and Promotion Project for Local and Up-scaling (SHEP-PLUS), a technical cooperation project between the Ministry of Agriculture, Livestock and Fisheries of Kenya and the Japan International Cooperation Agency. He worked for FRG II between 2010 and 2015 as an expert in charge of training and appropriate technology development.

Terutaka Niide is a consultant specializing in extension entomology and integrated pest management at RECS International. He worked on the FRG II project as an appropriate technology development expert.

Shingo Takeda was employed as an expert in the FRG II project.

Kiyoshi Shiratori is a consultant specializing in rural development at Kaihastu Management Consulting. He is a visiting professor at the Graduate School of Asian and Africa Area Studies, Kyoto University. He was a Chief Advisor in the FRG I and FRG II projects in Ethiopia.

Yoshiaki Nishikawa is a professor of Agricultural Economics and Resource Management at Ryukoku University, Japan. His main research theme is seed system management in Japan and African countries as well as participatory research.

References

Belias, D. and Koustelios, A. (2014) 'Organizational Culture and Job Satisfaction: A Review', *International Review of Management and Marketing*. 4(2), pp. 132-149.

NAS (National Academy of Sciences) (2009) *On Being a Scientist: A Guide to Responsible Conduct in Research: Third Edition*. The National Academies Press, Washington, DC.

Tharp, Bruce M. (2005) "Defining 'Culture' and 'Organizational Culture': From Anthropology to the Office." Industry White Paper. Haworth, Inc.: Holland, MI USA. <http://ap.haworth.com/docs/default-source/white-papers/defining-culture-and-organizationa-culture_51-pdf-28527.pdf?sfvrsn=6> [accessed 14 Dec 2014].

CHAPTER 12

Applying the FRG approach in agricultural extension: lessons from the Farmer Research and Extension Group approach

Belay Kassa and Dawit Alemu

Abstract

The chapter presents the experience, benefits and lessons learned from the application of the FRG approach as one tool of agricultural extension, as implemented by the Rural Capacity Building Project supported by the World Bank from 2006 to 2012. The project applied the Farmer Research and Extension Group (FREG) approach, which emerged from the national guidelines developed through adapting the FRG guidelines. The key lessons to ensure the suitability and scaling up of the approach are: 1) empowering the Offices of Agriculture at different levels to consider the FREG approach as one of the mechanisms of technology transfer; 2) empowering farmers to engage in agricultural experimentation; 3) fostering group action and assertiveness to make sure the offices concerned continue to support and supervise the FREGs; 4) empowering FREGs themselves to serve as sources of technology; and 5) creating and nurturing linkages with relevant stakeholders for improved access to markets.

Keywords: Farmer Research Groups, Farmer Research and Extension Groups, Rural Capacity Building Project, technology transfer, scaling up

Introduction

Though extension package testing was carried out through on-farm research by the research system previously, Farming Systems Research was promoted in the 1980s and represents the start of participatory research in Ethiopia, where the value of farmers' participation was fully recognized (Abera and Fasil, 2005). This linked with national recognition that technology development can only lead to agricultural innovation through strong multi-stakeholder participation. Accordingly, various farmers' participatory approaches have been adopted and have gradually moved from the consultative mode of participation in agricultural innovation systems to a collaborative mode of farmer participation, targeted to ensure 1) farmer empowerment, 2) improved technology adoption, and 3) better exploitation of indigenous knowledge. Participatory approaches have resulted in the involvement of agricultural research and extension organizations at all levels in farmer-demand identification, through, for example, the Farming Systems Approach (FSA), and Participatory Technology Development (PTD). In addition, participatory research has also been implemented through various forms of farmer groups (e.g. Farmer Research Groups (FRGs), Farmer Research and Extension Groups (FREGs), Farmer Extension Groups (FEGs), Farmer Field Schools (FFSs) and through approaches such as Participatory Learning and Action Research (PLAR) (Sanginga et al., 2006; Probst et al., 2003).

In the last decade, the members of the national agricultural research system (NARS) of Ethiopia have been promoting participatory research through FRGs, in which a group of farmers, extension workers and a multidisciplinary research team jointly participate in agricultural technology generation, verification, and improvement so as to meet farmers' needs and improve farmers' production and management practices (Endalkachew, 2008; Teklu, 2007). Following the implementation of the Rural Capacity Building Project (RCBP), the agricultural extension system has also been promoting the approach under the FREG name.

This chapter presents an assessment of the performance of FREGs as promoted through the RCBP, along with the lessons learned in adapting the FRG approaches in the area of agricultural extension. Specifically, it documents: (i) the characteristics of FREGs in terms of investments and technologies, outputs and activities and benefits generated, leading to a cost-benefit analysis, (ii) the lessons learned and the reasons for success or poor performance of FREGs, and (iii) the institutional arrangements for the sustainability and development of FREGs to the level of cooperatives or unions (as the next level of FREGs is expected to be formally organized as cooperatives).

Methods

The study was conducted by a team of researchers from the Extension Department of Haramaya University and the Agricultural Economics, Research-Extension and Farmers' Linkage Research Programme of the Ethiopian Institute of Agricultural Research (EIAR). It employed both formal and informal approaches to generate primary data. Focus Group Discussions (FGDs) and Key Informant Interviews (KIIs) were employed to supplement the primary data collected through two types of questionnaires that were tested prior to the survey. The first questionnaire was designed to generate data from development agents (DAs)[1] and the second one was used to collect data from FREG member farmers. Secondary data sources were also consulted. The data were generated in 2012.

The focus of the study was the RCBP intervention zones and *woredas*[2]. According to the records obtained from RCBP, the project was implemented in 131 target *woredas* that are spread across the country. More specifically, the 131 target *woredas* break down to 20 in Tigray, 23 in Amhara, 36 in Oromia, 29 in SNNP, seven in Afar, seven in Somali, two in Gambella, four in Benishagul Gumuz, two in Harari and one in Dire Dawa administrative regions.

As a sample for this study, about 20% of the target *woredas* were selected taking into consideration: 1) the existence of a national or regional research centre to enable data collection, 2) representation: representing remoteness and nearness to regional and zonal headquarters, and 3) accessibility. The total number of *woredas* covered by the FREGs evaluation study was 28 spread across the different regions of the country. The distribution of respondents (farmers and DAs) for the evaluation study by region, zone and *woreda* is presented in Table 12.1. The pre-tested questionnaires were administered to 102 farmer respondents and 51 extension workers.

In addition to the questionnaire-based data collection, FGDs were undertaken in all the sample *woredas* in the presence of experts from the *woreda* office of agriculture and with FRG farmers. Similarly, KIIs were conducted with researchers in the respective research centres of both regional and federal research institutes.

Approaches and institutional arrangements in promoting FREGs

The steps taken in promoting FREGs throughout the project were based on the national guidelines, developed from the FRG guidelines. The steps are: 1) undertaking a situation analysis and identification of needs, 2) selection of FREG member farmers, 3) selection of FREG leader/management, 4) activity planning, 5) implementation of planned activities, 6) capacity building, 7) participatory monitoring and evaluation, and 8) sharing experiences with other FREGs and farmers.

The results of the FGDs in the four regions show that selection of target groups and the formation of a FREG, objective identification, partnerships formation and location identification were undertaken after sensitization workshops at different levels (zonal, *woreda*, and *kebele*[3]), organized by the project. In general, the FREGs were established following two approaches: (i) the project together with Bureau of Agriculture (BoA) staff undertaking the responsibility of establishing and running FREGs, mainly through appointing focal persons at zonal and *woreda* levels, and (ii) the project delegating agricultural research centres to promote FREGs. In the first approach the fund was managed by the BoA for the different levels and in the second case, the fund was transferred to and managed by the respective research centres.

Based on the FGD made in the selected *woredas*, several challenges were observed. The key challenges for the project and BoA staff-managed FREGs were: 1) inadequate supervision and follow-up of established FREGs mainly owing to the limited mobility of the focal persons, 2) limited skills of DAs

168 FARMER RESEARCH GROUPS

Table 12.1 Distribution of respondents by region, zone and *woreda*

Region	Zone	Woreda	No. of farmers	No. of DAs and SMSs	Focal Research Centre(s)
Tigray	Southern Tigray	Ofla	4	3	Humera RC, TARI
Amhara	West Gojjam	Bahirdar Zuria	5	2	Adet RC, ARARI,
		Yilmana Densa Adet	6	2	Bahir Dar University Adet RC
	North Shewa	Menjar Shenkora	-	2	Bebre Berhan RC, ARARI
		Basona Worana	5	-	Debre Berhan University
Oromia	East Harerge	Gurewa	3	1	Haramaya University
	West Shoa	Ambo	-	4	Ambo RC, EIAR
		Holeta	10	2	Holeta RC
	East Shoa	Adama	9	5	Melkassa RC, EIAR
		Dugda	5	2	Debre Zeit RC, EIAR
	S/W/ Shoa	Weliso	6	1	Holeta RC, EIAR
SNNP	Sidama	Dara	4	-	Hawasa University
	Wolaita	Damotgale	5	4	Sodo University
Afar	Zone 3	Amibara	5	10	Werer RC, EIAR
		Daifagie	3	-	Werer RC, EIAR
Somali	Jigjiga	Jigjiga	3	2	Jigjiga RC, SOPARI
		Gursum	2	1	Haramaya University
Gambella	Agnuak	Abobo	5	5	Abobo RC, GARI
Benishan gul Gumuz	Metekel	Dangur	5	2	Pawe RC, EIAR
		Pawe	5	2	Pawe RC, EIAR
Harari	Harari	Erer	4	1	Haramaya University
Diredawa	Diredawa	Diredawa	8	-	Haramaya University
Total no. of respondents			102	51	

Note: - indicates that there was no respondents

RC: research centre, TARI: Tigray Agricultural Research Institute, ARARI: Amhara Regional Agricultural Research Centre, EIAR: Ethiopian Institute of Agricultural Research, SOPARI: Somali Pastoral and Agro-pastoral Research Centre, GARO: Gambella Agricultural Research Institute

and focal persons about the principles and purpose of FREGs, and 3) the limited possibility of sustaining the smooth functioning of the established FREGs. For FREGs established by research centres, the key challenges were: 1) limited linkage with extension for scaling up the approach together with any technologies/innovation, and 2) the support being focused on technology provision and adaptation with limited emphasis on other aspects such as marketing.

Characteristics of FREGs by type and region

The RCBP has promoted a total of 651 FREGs in all the regions of the country. The average number of member farmers per FREG is around 20. However, this average masks regional differences which vary from 10 in Gambella and Harari regions to 36 in the Somali region.

The mean difference in the number of member farmers across the regions was not statistically significant. Similarly, the average proportion of women farmers (female-headed households) per FREG was 18%, with a significant difference across regions ranging from zero in Somali and Harari regions to 38% in Tigray region (although the national gender mainstreaming guideline in the agricultural sector states a minimum of 30% participation of female farmers) (MoA, 2011) (Table 12.2).

The established FREGs were engaged in different agricultural activities. More specifically, of the 651 FREGs, 292, 145, 105, 26, 19, 17, and eight were engaged in activities related to field crops, horticultural crops, livestock (cattle, shoats), forage, poultry, bee keeping and post-harvest management,

Table 12.2 Average number of member farmers and proportion of women members in FREGs, by region

Region	Total number of FREGs	Average number of members per FREG (SD)	Average proportion of females in a FREG (SD)
Tigray	47	13 (8)	0.38 (0.12)
Afar	45	13 (6)	0.08 (0.06)
Amara	117	23 (5)	0.19 (0.11)
Oromia	87	20 (34)	0.26 (0.32)
Somali	30	36 (22)	-
Benishangul-Gumuz	43	25 (22)	0.10 (0.11)
Gambella	13	10 (6)	0.30
SNNPR	254	16 (6)	0.14 (0.14)
Harari	8	10 (0)	-
Dire Dawa	7	26 (0)	0.15
Total	651	20 (21)	0.18 (0.22)
Mean difference across region (T-value)		0.89	2.47**

Note: Figures in parentheses indicate the standard deviations

** significant at 5% probability level

Table 12.3 Distribution of FREGs by region and activity

FREG Regions	Tigray	Amhara	Gambella	Dire Dawa	Afar	Benishangul-Gumuz	Oromia	SNNP	Harari	Somali	Total
Crops	15	58	9	-	3	10	18	168	-	11	292
Horticulture	14	25	4	3	14	27	24	18	5	11	145
Livestock (Ruminants)	11	20	-	2	13	2	40	7	2	8	105
Forage	2	2	-	-	8	-	-	14	-	-	26
Poultry	-	1	-	2	-	1	-	14	1	-	19
Bee keeping	4	3	-	-	2	-	-	8	-	-	17
Post-harvest	-	3	-	-	-	-	5	-	-	-	8
NRM	-	3	-	-	-	-	-	-	-	-	3
Others	1	2	-	-	5	3	-	25	-	-	36
Total	47	117	13	7	45	43	87	254	8	30	651

Source: RCPB, 2012

respectively. Only three FREGs were involved in natural resource management activities, while the remaining FREGs (36) were engaged in various other activities. The distribution of the 651 FREGs by activity type and region is presented in Table 12.3.

Member farmers' perceived involvement in FREG establishment and governance

The survey results indicate that FREGs are normally established through the initiation of DAs, local administrators and researchers. Only 13% of the FREG member farmers reported that they became members through their own initiative (Table 12.4). However, most of the member farmers (80%) indicated that they were aware of the criteria for membership, which include: 1) similarity in terms of interest and preference, 2) proximity to each other, 3) gender composition, and 4) willingness to be a member.

For the effective functioning of FREGs, the commitment of all members is very important. In response to a question regarding the perception on the level of commitment of fellow member farmers, 5% and 13% of the respondents judged it as very poor and poor, respectively. However, 51% and 31% of the respondent farmers rated the level of commitment of other member farmers

Table 12.4 Member farmers' responses on FREG membership and associated issues

Indicator		% of member farmers
How the farmer became a member	Selected by DA	51
	Selected by local administrator	21
	Selected by DA in consultation with local administrator	12
	Farmer asked to be a member	13
	Selected by researchers	4
% of farmers aware of the selection criteria to be a member	Yes	80
Respondents' rating about the commitment of other FREG members	Very poor	5
	Poor	13
	Good	51
	Very good	31
Happiness of respondents about all members of FREG	Yes	87
% of members reporting that enough support was provided by researchers and/or extensionists	Yes	58
% of member farmers who were trained on FREG or related issues	Yes	86

Note: n = 102

as good and very good, respectively. Cordial and harmonious relationships among members is one of the important preconditions for the success of a FREG. In this respect, 87% of the member respondents reported that they had excellent relationships with other member farmers (Table 12.4).

In general, FREGs require technical and managerial support from both researchers and extensionists[4] (DAs and SMSs). However, only 58% of the respondent member farmers believed that the support provided by researchers and extensionists was adequate. In terms of training on FREGs and related issues, a great majority of the respondent member farmers (86%) reported their participation in different training programmes.

One of the key principles of a FREG is the promotion of farmers' participation in the implementation of its activities along with involvement in decision-making in the process. The guideline for members' participation indicates full participation. However, the survey results presented in Table 12.5 show clearly that a significant proportion of the respondents perceived their involvement in FREG-related decision-making as marginal.

Similarly, one of the important principles in the implementation of FREG activities is the joint evaluation of the selected activities (by farmers, DAs and

Table 12.5 Ratings of member farmers about their involvement in FREG-related decisions

Decisions	Rating (% of respondents)			
	Very high	High	Low	Very low
FREG leader selection	27	31	23	19
Commodity selection	25	31	23	21
Site selection	37	34	20	9
Land preparation	44	37	13	6
Sowing	43	38	13	6
Experimental activity	28	23	23	26
Evaluation	29	25	20	25

Table 12.6 Average number of joint evaluations made on the performance of field experiments (farmers, DAs and researchers)

Region	Mean	Standard deviation
Tigray	1.00	1.73
Afar	2.00	0.93
Amhara	6.50	3.68
Oromia	3.55	6.22
Somali	0.00	0.00
Benishangul-Gumuz	3.10	1.45
Gambella	1.40	0.89
SNNPR	6.25	7.19
Harari	0.00	0.00
Dire Dawa	1.00	0.53
Total	3.19	4.55
Mean difference (T-value)	2.33**	

researchers) to better facilitate experimentation and the learning process. The survey results reveal that the number of joint evaluations made, as reported by the respondents, varied considerably across regions. A closer look at Table 12.6 shows that while there was no joint evaluation in the Harari and Somali regions, the maximum number of joint evaluations (6.5) was reported for the Amhara National Regional State.

Performance of FREGS: cost-benefit analysis

Perceived benefits of FREG membership

Both FREG member farmers and FREG promoters (DAs and SMSs) were asked about the observed performance and perceptible benefits of FREGs and their responses are summarized in Table 12.7. The results presented in the table indicate that the majority of the respondent member farmers and FREG promoters believed that the established FREGs brought about tangible benefits to their farming communities. More precisely, they perceived that the established FREGs (i) increased the use of improved technologies, (ii) improved farmers' skills in undertaking simple agricultural experiments, (iii) improved the capacity of farmers to work with DAs, SMSs and other farmers, (iv) improved farmers' capacity to access other agricultural inputs; and (v) improved farmers' access to markets.

Table 12.7 Perceived benefits of FREG to member farmers

Respondents	Perceived benefits	Ratings (% of respondents)			
		Very poor	Poor	Good	Very good
FREG members (n=102)	Increased use of improved technologies	2	7	48	43
	Improved skill in experimentation	12	25	34	28
	Improved capacity to access agricultural inputs	3	17	54	26
	Improved capacity to access market	12	19	47	23
	Improved capacity to work with DAs and SMSs	2	13	50	35
	Improved capacity to work with other farmers	2	13	52	33
DAs and SMSs (n=51)	Increased use of improved technologies	2	10	40	48
	Improved skill in experimentation	-	18	50	32
	Improved capacity to access agricultural inputs	2	16	44	38
	Improved capacity to access market	10	20	39	31
	Improved capacity to work with DAs and SMSs	2	8	46	44
	Improved capacity to work with other farmers	4	8	44	44

It should also be noted that the results presented in Table 12.7 show that a considerable proportion of the member farmers as well as DAs and SMSs categorized some of the perceived benefits of FREGs in the very poor and poor categories. The reasons for the unfavourable judgement as elicited in the FGDs and KIIs include: 1) the concern that FREGs had not received proper support and supervision, 2) the lack of appropriate activity/technology, and 3) the limited skill of the promoters, especially the focal persons. For example, members of the FREG working on wheat in Ofla *woreda* of the Southern zone in Tigray region reported that in the first year of the activity most of the farmers were abandoning membership owing to the disease susceptibility of the introduced new wheat variety. However, following a change of variety by Alamata Agricultural Research Centre of Tigray Agricultural Research Institute and strong follow-up, other farmers were willing to join. Similarly, as indicated above, about 55% of the respondent DAs and SMSs reported that the support provided to the established FREGs was inadequate mainly owing to the limited type and amount of technologies supplied (84%), the inadequacy of the training given to the farmers (67%), and inadequate supervision and follow-ups (62%).

The key factors contributing to the poor performance of some of the FREGs as elicited during the FGDs are discussed below.

Limited support and supervision. One of the key requirements for the proper functioning of FREGs is the provision of adequate technical support pertaining to the management/governance of FREGs and technical skills in technology adaption and experimentation. Equally important is regular supervision to follow the progress made and take quick remedial measures in cases where there are emerging challenges related to the technology itself or to FREG governance. The results from the FGD and KII indicated that most of the FREGs lacked adequate support and supervision. A typical case is that of the FREGs at Erer *woreda* of Harari region, where all the FREGs were not practically functioning (Box 12.1). On the other hand, with different challenges, the goat FREGs at Grawa Woralo-Busa *kebele*, Dire Dawa Administration were successful mainly owing to the timely supervision and provision of required support to the member farmers when needed (Box 12.2).

Limited skill of promoters, especially the focal persons. The FREGs were promoted through the assignment of focal persons at different levels. The results from FGD and KII indicated that most of the focal persons lacked adequate knowledge about FREG (principles, approaches, governance etc.) and were not knowledgeable about the technologies. This has been one of the main causes of the limited technical support provided to most of the FREGs.

Lack of appropriate activity/technology. This issue is in a way related the challenges discussed above. The results of this study revealed that there were a considerable number of FREGs that selected and promoted technologies that

> **Box 12.1 FREGs in Erer *woreda*, Harari National Regional State (FGD result)**
>
> FREGs were established in 2010 to work on cereal crops through the Erer *woreda* Bureau of Agriculture in collaboration with Haramaya University. The member farmers reported that most introduced varieties of the different cereal crops were susceptible to diseases and insects. Therefore they were producing less than the established local varieties. Following the poor performance of the activities undertaken in the FRGs, nearly all the FREGs were abandoned and currently there is no FREG functioning in the *woreda*. The main reasons for the failure of the FREG activities were reported to be: (i) the limited responses of the researchers to assist farmers and promoters to address the disease and insect problems, (ii) inadequate skill and capacity of local promoters, especially of DAs, (iii) the lack of adequate training to DAs and member farmers, and (iv) lack of technical support and adequate supervision.

> **Box 12.2 Success story of goat FREG at Grawa Woralo-Busa *kebele*, Dire Dawa Administration**
>
> The goat FREG was promoted by the Dire Dawa BoA and Haramaya University in 2010. In the selected *kebele*, FREG member farmers were selected by DAs and the selection criteria used included resource status and level of interest. Unlike other areas where resource-rich people are selected for membership, in this *kebele* resource-poor farmers were selected for FREG membership. A total of 27 farmers were selected to be members of the goat FREG. The primary objective of the FREG was to support resource-poor farmers and help them improve their livelihoods, while at the same time encouraging the process of knowledge generation and experimentation.
>
> As the newly established FREG was to focus on goat fattening, improved breeds were procured from Haramaya University with the financial support of the Dire Dawa BoA. While training on the management of goats, feeding practices and reproduction aspects was provided jointly by Haramaya University and BoA, member farmers were responsible for the proper management of the goats. Haramaya University and the Dire Dawa BoA also provided the necessary inputs. The FREG members reported that they benefited from close and regular supervision of a team of experts drawn from the Dire Dawa BoA (DAs and SMSs) and Haramaya University.
>
> The FREG member farmers reported that the FREG approach was suitable for their circumstances and the partnership established were useful. Moreover, they pointed out that the interactions and exchange of information among member farmers were very good. The FREG members reported also that following the success registered by the goat FREG, other FREG projects with different goals had been established. At the time of the survey (May 2012), it was apparent that a scaling-up process was under way, as many farmers were interested in participating in the FREG approach.

were not appropriate to the area and/or were engaged in activities that were not appropriate. When this occurred in areas where there were good practices of timely supervision and support, the necessary adjustments were made in the same year or in the subsequent year. However, in areas where there was no properly organized support and supervision, farmers withdrew from FREGs. These problems were more prevalent in areas where FREGs were promoted without a strong link with an agricultural research centre. Cases in point include the cereal crop FREGs established in Erer *woreda* of the Harari region, where the local variety performed much better than the introduced varieties. In this particular case, as there was no proper supervision and follow-up, it was not possible to address the problem in time.

Costs of FREGs

The costs of FREGs are estimated based on the budget allocated and the number of FREGs established and supported by the project. The FGDs and discussions with the RCBP staff members revealed that the established FREGs were supported for three years. The type and level of support, including the amount of money allocated, varied by FREG type and region. However, the average cost per FREG is estimated by dividing the amount of utilized budget by the actual number of FREGs supported each year. It must be noted here that as each FREG received financial support from the project for a period of three years, the average costs are estimated by considering only those FREGs that received financial support from the project during the years under consideration.

The estimated cost of support for each FREG during the project period is summarized in Table 12.8 both in terms of the amount allocated and utilized year by year. The figures show that in some years there was underutilization and in others there was overutilization. The amount of budget allocated and utilized shows a decreasing trend in the three years from 2008/09 to 2010/11, associated with the fact that established FREGs are supported for three years and in recent years the number of FREGs has decreased.

The result indicates that on average, 13,189.12 Ethiopian birr was spent per FREG every year. Taking into consideration the average number of member farmers in each FREG (20), the cost incurred for FREG-related activities per member farmer per year was about ETB 659.46. This estimate does not consider the indirect cost that the member farmers and extension workers incurred. This is mainly considering an assumption that these indirect costs are sunk costs, which are costs that are incurred and cannot be recovered by any means.

Table 12.8 Budget allocated by the RCBC for FREGs and related activities and its utilization

Budget year	Allocated budget (ETB)	Utilized budget (ETB)	Proportion of budget utilized (%)	Average cost per FREG (ETB)
2007/08	728,146.30	728,146.30	100	8,823.76
2008/09	15,281,700.00	6,204,140.60	41	21,145.06
2009/10	15,883,000.00	8,385,037.58	53	18,663.21
2010/11	9,461,130.70	8,806,824.96	93	13,920.27
2011/12	2,058,700.00	2,601,648.18	126	4,576.51
Total	43,412,677	26,725,797.62	62	13,189.12

Source: RCBP Office

Additional benefits of membership

Though the FREGs provide different benefits to member farmers and other community members, some of these cannot be estimated in monetary terms. Consequently, in this study the cost-benefit analysis of the FREG intervention is made by narrowing down the benefits to the possible effects of using improved agricultural technologies only. Moreover, so that we can see the financial benefits accruing to member farmers as a result of their participation in FREGs, the cost-benefit analysis is done for a typical farmer who produces OPV maize varieties, which have been used in the FREG activities (varieties Melkassa-2 and Melkassa-4).

Based on the KII with FREG member farmers, a typical member farmer is considered to have the following characteristics:

1) allocates 0.25 ha of his land for maize;
2) used to achieving the national average yield of maize (20 quintals/ha);
3) through engagement in a FREG, the farmer adopted improved varieties of maize, which increased the level of productivity in the range of 38–48 quintals/ha and the typical farmer gets an average yield of 43 quintals/ha;
4) influences other farmers in adopting the maize varieties equally in the range of five to 10 farmers. For the purpose of this analysis it is assumed that a typical member farmer influences on average seven other farmers.

Accordingly, the typical FREG member farmers and other farmers who adopted the maize varieties got a yield advantage of 23 quintals/ha. Since the typical farmer owns one-quarter of a hectare for maize production, the actual yield gain per member is (on average) 5.75 quintals. This implies that the intervention through FREG brought a yield gain of 40.25 quintals under the assumption that a typical farmer influences seven farmers to adopt the technology. In financial terms, taking into consideration the price of maize estimated at ETB 400 per quintal, the typical FREG member received an income advantage of ETB 16,100.00 on average per year from the average cost of investment estimated at ETB 659.46 per year per member.

Sustainability and exit strategy

The operation of the established FREGs is more or less dependent on the continuation of project funding and administration, therefore most of the FREGs would not be able to operate after support from the project ceases. The case in Ofla *woreda* of Tigray shows that the established FREGs are not operating any more, even though the activities resulted in increased demand for technologies (Box 12.3). Some FREGs that are strategically linked with research centres and markets seem to continue their operation even after the project support is withdrawn. For example, the six FREGs promoted by the Debre Zeit Agricultural Research Centre in collaboration with the *woreda* BoA at Memihir Agar *kebele* of Minjar Shenkora *woreda*, North Shewa zone, Amhara

> **Box 12.3 A wheat FREG at Ofla *woreda*, Southern zone, Tigray region**
>
> The FREG approach was considered an important mechanism in evaluating and disseminating new technologies. The Alamata Agricultural Research Centre in collaboration with the Office of Agriculture of the *woreda* and the zonal BoA established a wheat FREG at Hashengie *kebele*. In the first year of operation, there was limited interaction between farmers, researchers and extensionists. Moreover, the varieties were susceptible to pest attack and disease infestation and the yield levels were no better than the local varieties. The following year, however, the research centre introduced disease-resistant wheat varieties with better yield levels. The newly introduced varieties were successful and this resulted in the increased interest of other farmers to join the FREG in subsequent years.
>
> In the course of the FGDs, farmers stated that they were ready and motivated to adopt new technologies. However, it is sad to learn that the wheat FREG in question was no longer operating, mainly owing to the termination of the engagement of the Alamata Agricultural Research Centre and the absence of a considered exit strategy that could have ensured the sustainability of the FREG.

> **Box 12.4 Success story of crop FREGs at Minjar Shenkora *woreda*, North Shewa zone, Amhara region**
>
> The FREG approach was promoted by the Debre Zeit Agricultural Research Centre in collaboration with the *woreda* BoA to focus on three major crops grown in the *woreda*: wheat, chickpea and lentil. The results of the FGDs at Memihir Agar *kebele* indicate that there were six FREGs, each consisting of 10 to 15 farmers. These FREGs are strategically supported by DAs and researchers to serve as the seed source for other farmers in the *woreda* in general and in the *kebele* in particular.
>
> The FREG member farmers unanimously reported that they greatly benefited from introduced varieties. The majority of farmers in the *woreda* are now aware of the success stories of these FREGs and are ready to form FREGs. Their motivation is manifested by the fact that many farmers are establishing groups and selecting sites by themselves. There is high demand in the *woreda* for seed of improved varieties and the FREGs are striving to fulfil the demand. The FREGs are also working on finding potential markets for their produce and have organized a field day which involved different stakeholders.
>
> The FREGs were found to embrace participatory approaches in that member farmers consistently reported that they were involved in all activities beginning with land preparation through to evaluation of outcomes. The Debre Zeit Agricultural Research Centre was praised by farmers for its continuous introduction of technologies, provision of training, monitoring, supervision, and support. The farmers stated that they were very happy about the direct and strong linkages established between farmers and researchers.

region, are operating successfully as they are serving as a seed source for other farmers within the *woreda* and other neighbouring *woredas* (Box 12.4).

These practical examples indicate that in order to ensure the sustainability of the FREGs, there is a need to design right from the very beginning an exit strategy that assures the smooth operation of the FREGs after support from the project is withdrawn. In this respect, possible strategies include: 1) empowering the BoA at different levels to consider the FREG approach as one of the mechanisms of technology transfer, empowering farmers to engage with agricultural experimentation also fostering group action and assertiveness so as to make sure that the offices concerned continue to support and supervise the FREGs (see Box 12.2 and 12.3); 2) empowering the FREGs themselves to serve as sources of technology (see Box 12.4); and 3) creating and nurturing linkages with relevant stakeholders (cooperatives, seed enterprises, traders, processors etc.) for improved access to markets. Though the longer-term outcome is yet to be seen, the decision to use FREGs as one extension approach by the BoA of the Finfine Zuria zone of Oromia region is exemplary in the institutionalization and sustainability of FREGs.

Lessons learned

In spite of the serious challenges encountered relating to limited support and supervision, lack of appropriate activity/technology, and the limited skills of promoters and especially of focal persons, the project-promoted FREGs have been instrumental in increasing the use of improved technologies and improving farmers' skills in agricultural experimentation. The experience of the few FREGs that have been promoted to micro-seed enterprises or those that have become seed suppliers to formal seed enterprises (Ethiopian Seed Enterprise, Oromia Seed Enterprise, South Seed Enterprise and Amhara Seed Enterprise) show that the following key points and lessons are important to sustain FREG activities:

- It is important to strengthen/establish linkages with agricultural research centre(s), seed enterprises, and/or cooperatives for improved technology transfer and marketing.
- The selection of technologies has to be made by involving all relevant actors. Moreover, the list of potential technologies that can be suitable for the target agro-ecologies has to be made available. This implies the need for undertaking a situation analysis and options-needs matching process with farmers.
- A reliable supply of technologies/seeds has to be assured before starting a FREG. Therefore, the activity selection for FREGs must consider criteria relating to the availability of technologies.
- There is a need to train focal persons and DAs not only about technologies but also about FREG principles and governance.
- The need to provide adequate support and regular supervision must be high on the agenda of FREG promoters and other concerned bodies.

Conclusion

Through promotion of the FREG approach, the project has played an important role in improving the use of agricultural technologies and in boosting the skills of farmers in agricultural experimentation, to varying degrees across regions. The overall performance of most of the FREGs is rated to be either 'good' or 'very good' by more than 80% of the study respondents. For example, in terms of increasing the use of improved technologies, 91% of the respondent member farmers rated the performance of FREGs as 'good' and 'very good'. Similarly, 88% of the respondent DAs and SMSs rated the performance of FREGs in increasing the use of improved technologies as 'good' and 'very good'. These results point to the fact that a higher proportion of the FREGs have performed very well.

The poor performance in some of the FREGs as reported is linked with poor support and supervision provided by FREG promoters; the lack of appropriate activities/technologies; and the limited skills of the promoters and particularly of the focal persons.

The cost-benefit analysis based on the assumption of a typical FREG member farmer indicates that the cost of promoting FREG-related activities per member farmer was ETB 659.46 per year and a member farmer was able to generate and help others generate financial benefit ranging from ETB 9,000 to 28,000 per year.

The lessons learned indicate that most of the established FREGs will have serious problems continuing operation once support from the project is withdrawn. There is an absence of a considered exit strategy which should have been in place from the initial stages of establishing FREGs. The different actors that have been involved in the promotion of FREGs need urgently to design a strategy to sustain the established FREGs once support is withdrawn. In this regard, the future application of the FRG approach in Ethiopia through FREGs or other participatory approaches needs to consider the following key issues:

- creating appropriate interfaces between FRG and FREGs to align activities and facilitate smooth communication;
- empowering the Offices of Agriculture at different levels to consider the FREG approach as one of the mechanisms of technology dissemination and popularization;
- empowering farmers to engage in agricultural experimentation and fostering group action and assertiveness so as to make sure that the concerned offices continue to support and supervise the FREGs;
- empowering the FREGs themselves to serve as sources of technology to local farmers;
- strengthening/establishing linkages between FREGs and relevant stakeholders (cooperatives, seed enterprises, traders, processors, etc.) for improved market access;
- exploring possibilities for institutionalizing the approach as a linkage/interface mechanism between research and development in the country.
- Further developing/refining the approaches followed in FREG promotion, including elucidating in detail the conceptual and practical differences between the FRG and FREG approaches so as to avert confusion among researchers and practitioners.

About the authors

Belay Kassa is Professor of Agricultural Economics and Interim Deputy Rector of Pan African University and African Union Commission, Addis Ababa. He previously was the President of Haramaya University.

Dawit Alemu is Director of the Agricultural Economics, Extension and Gender Research Directorate of the Ethiopian Institute for Agricultural Research. He has been associated with EIAR since 1999 as a Senior Researcher and Coordinator. His research focuses on agricultural marketing with an emphasis on agricultural inputs.

Notes

1. Development agents serve as frontline extension workers directly working with farmers.
2. *Woreda* is an administrative division in Ethiopia equivalent to a district.
2. A *kebele* is an administrative division in Ethiopia equivalent to a village.
3. Extensionists are those working for the Ministry of Agriculture and include DAs and subject matter specialists (SMSs). SMSs serve as extensionists providing training and technical backstopping to DAs.

References

Aberra D. and Fasil K. (2005) 'An Overview of Participatory Research Experience in Ethiopian Agricultural Research System', in proceedings of a workshop on farmer research group (FRG): concept and practices. Addis Ababa, Ethiopia

Endalkachew Z. (2008) 'Evaluation of Farmer Research Extension Group (FREG) as Extension Approach: The Experience of Sida-Amhara Rural Development Program in Kalu District, of Amhara Region, Ethiopia', MSc thesis, Laurenstein University of Applied Sciences, The Netherlands.

Ministry of Agriculture (2011) 'Guidelines for Gender Mainstreaming in the Agricultural Sector', Women's Affairs Directorate, Ministry of Agriculture, Addis Ababa, Ethiopia.

Probst, K., Hagmann, J., Fernandez, M. and Asby, J.A. (2003) *Understanding participatory research in the context of natural resource management: paradigms, approaches and typologies*, Agricultural Research and Extension Network UK, The Overseas Development Institute, 111 Westminster Bridge Road, London SE1 7JD, UK.

Sanginga, P., Tumwine, J. and Lilja, N. (2006) 'Patterns of participation in farmers' research groups: lessons from the highlands of south-western Uganda'. *Agriculture and Human Values*, 23(4), pp. 501-512.

Teklu, T. (2007) 'Powerless when acting in isolation but superpower when acting together: farmer research groups (FRGs) and research extension advisory councils (REACs) as mechanisms that enhance concerted effort among the actors of agricultural development', Ethiopian Institute of

Agricultural Research, Research-Extension-Farmer Linkages Department, Mimeograph, Addis Ababa.

RCBP (Rural Capacity Building Project). 2012. Implementation completion and results report (IDA-42010 TF-90084). Report No: ICR2539. The World Bank, Washington DC. USA.

PART IV
Conclusion

CHAPTER 13

Conclusion: Recommendations for strengthening the responsiveness of agricultural research systems

Dawit Alemu, Kiyoshi Shiratori, Taku Seo and Yoshiaki Nishikawa

Abstract

This chapter describes conclusive findings from the experience of two Farmer Research Group (FRG) projects primarily within the context of the Ethiopian national agricultural research system (NARS). It also describes agriculture development policies and implementation processes with reference to recent global discussions on participatory agricultural research (PAR). For this purpose, a brief summary of debates on PAR in general and its application and implications in Ethiopia are discussed first. Popular examples of system and network approaches for innovation and rural development are explained and then the reality of the Ethiopian case is described, based on 10 years' FRG experience and the findings described in previous chapters. A summary of findings is presented for use in the further application of FRGs in Ethiopia, including a slightly different idea of institutionalization, which is a simultaneous process of implementation and scaling up within the NARS system, rather than model development and scaling up in chronological order. Finally, the chapter gives recommendations for further study.

Keywords: degree of participation, Ethiopia, Farmer Research Group, innovation, institutionalization, participatory agricultural research, researchers

Introduction

This book has presented the experiences gained while promoting the participatory approach through Farmer Research Groups (FRGs) based on two projects implemented in 2004-2015 by the Ethiopian Institute of Agricultural Research (EIAR) in collaboration with the Japan International Cooperation Agency (JICA). The Project on Strengthening Technology Development, Verification, Transfer and Adoption through Farmer Research Groups (FRG I) piloted the approach with the Melkassa Agricultural Research Centre (MARC) of EIAR and the Adami Tulu Agricultural Research Centre (ATARC) of the Oromia Agricultural Research Institute (OARI) in the Central Rift Valley.

http://dx.doi.org/10.3362/9781780449005.013

The Project for Enhancing Development and Dissemination of Agricultural Innovations through Farmer Research Groups (FRG II) targeted the promotion of the approach within the country's national agricultural research system (NARS), which includes EIAR, regional agricultural research institutes (RARIs) and universities with agricultural faculties.

The FRG approach is characterized by group/collective action, multidisciplinary teams and on-farm trials with the research focus on farmers' needs and outputs that are technically, economically and socially feasible by and for farmers. In every step of its application, the FRG approach activities are jointly implemented with clear responsibilities for researchers, farmers, extension workers and other stakeholders in each stage through planning, data collection, monitoring, analysis and information sharing. This allows the involvement of more participation by researchers as an essential component as well as farmers demonstrating ownership of both the process and the outputs of the research activities. It was with this premise that the two projects promoted the FRG approach within the NARS and there have been considerable achievements both in terms of the wider application of the approach and also the research outputs.

In this chapter, conclusive findings from the experiences of FRG projects are summarized primarily within the context of Ethiopian agricultural development, but with reference to the global discussion on participatory agricultural research (PAR). A brief summary of the debates around PAR and its application and implications in Ethiopia are discussed first. Actual findings are presented for the further application of FRG in Ethiopia and further afield, particularly in Sub-Saharan African countries.

Debates on participatory agricultural research in development

Participatory agricultural research started in various ways, but the most common understanding was that conventional research, based at research stations, was not responsive to the actual needs of the farmers for whom the various technologies were being designed. PAR has been much talked about, and implemented for the last 30 years or so. Huge numbers of reports have been produced and it even seemed as though it had become the mainstream of agricultural research at one stage, particularly in the 1990s. The term 'participatory agricultural research' has many meanings and implications. Okali et al., (1994) analysed different types and usages of 'participatory research', since there were so many variations. Furthermore, the term 'participatory approach' has even more diverse implications. One typical set of statements associated with participatory agricultural research presents participation as a mechanism to increase efficiency and effectiveness; participation as a fundamental right and a process in which community members and other citizens mobilize for collective action, empowerment, institution building, inclusive deliberation and politically negotiated processes; and even that citizens are seen as clients or consumers and are asked to participate by paying for goods and services provided by the market, in so doing becoming more economically efficient actors (Pimbert, 2004). Pimbert cautioned that care should be taken in the wide application of the term 'participatory research', since this could lead to widespread disillusionment with people-centred approaches and result in the discrediting of the very concept of participation.

However, recent discussions on PAR are more about system and network aspects of participation and innovation. For example, Neef and Neubert (2011) present participatory research as a system consisting of six dimensions and 30 attributes. They state that 'these six dimensions and the related attributes are intended to cover the main parameters needed to describe the participatory elements employed in a given project in a systematic way', with the purpose of 'optimizing' participatory methods, rather than maximizing them.

One of the leading institutes in participatory research, the International Institute for Environment and Development (IIED), has published many research papers and manuals regarding the meaning of and tools for participatory research. Pimbert (2012) suggests that participatory research is regarded not as a scientific research field but as 'a power game field among players'. In his paper, democratization means farmers' participation in agricultural research. More participation in more phases of research needs to be realized. The more farmers are sought, the more upstream choices and designing of scientific innovation (e.g. planning of research) farmers can participate in. This is similar to the idea of degree of participation, which is usually explained as starting with passive participation, and then improving to participation in information giving, participation by consultation, participation for material incentives, functional participation, or interactive participation, and finally reaching self-mobilization (Pretty et al., 1995: 61). The statement by Pimbert seems different from this idea of degrees of participation, since he mentions that innovation can mean simply the acceptance of farmers' suggestions by high-level policy makers. In this way, 'participatory research' has a tendency to depart from farmers' fields and also from the researchers working with the farmers on those fields.

Some authors, such as Li et al., (2013), argue about the technical aspects of participatory research following their analyses on two different methods of participatory breeding, namely participatory plant breeding of hybrid varieties and open pollinated varieties (OPVs). They conclude that OPV development allowed more 'collegial participation', whereas participatory hybrid variety breeding 'allowed a form of collaboration in which breeders and farmers shared tasks, along lines determined by the formal research institute'. The main purpose of participatory plant breeding (PPB) is not about the political aspects of participation, but technical improvement and developing knowledge of both farmers and breeders. That is, PPB of hybrid varieties can improve farmers' knowledge about how to breed hybrid varieties by themselves. On the other hand, OPVs can increase breeders' knowledge about local conditions and so on. Therefore, the stage at which farmers can participate should be decided based not on political rights, but on their comparative advantages.

Wageningen University has various research projects and educational courses related to participatory research. Among them, Participatory Approaches in Agricultural (technology) Development and their Up-scaling Programme for post-graduate training emphasize the participatory attitudes of researchers from developing countries (Almekinders et al., 2009). Their emphasis is on the competencies researchers should be equipped with and how these competenices could be developed. It is worth noting their efforts in real situations and workshops with time for reflecting on how to develop such skills and attitudes. Basic scientific literacy is noted by the authors as a separate matter from participatory skills, however, and scientific literacy seemed to be taken for granted for those PhD candidates. Therefore, a combination of scientific skills and political and/or communication skills and attitudes are discussed as different matters for institutionalizing participatory research.

Blaakman (2003) proposed the importance of 'unlearning' the educational discipline that facilitators had obtained during formal education. She discussed the belief of extension workers that they are responsible for giving advice to solve others' problems. This attitude may originate from the deficit model of education in which farmers lack capacity owing to poor education and learning opportunities, therefore those who have more education assume the need to teach them. Since a failure to see farmers as equal partners is one of the major obstacles to successful participatory research, this is an important consideration. Self-reflection assisted by others acting as role models is beneficial, and this needs to be institutionalized in higher education if participatory approaches are going to be mainstreamed in research (Hagmann and Almekinders, 2003). Some of the research carried out by PhD projects at other Wageningen programmes follow the trend of system and network approaches. One example of potential relevance to the Ethiopian context is from a case in Bolivia. Blandon (2014) analysed different organizations working on participation in Bolivia and found that participation designed by technically oriented non-government organizations, such as PROINPA, is not just 'technical', as some professionals would like to perceive it, or simply 'political' as expressed in some critical views on participation. Instead, it is 'malleable' in the sense that each actor is involved in finding new balances between technical, economic, and political considerations. As this thesis tries to illustrate, this is true for both PROINPA's technicians and the politicized interventions of another NGO working in the area, EMAPA. Blandon's thesis approached participation from the macro-level, which means participation in national political decisions and participation in markets are referred to together with more technically oriented activities at a micro-project level. More and more NGOs and donors tend to adopt this model for scaling up participation.

Another interesting output from the Wageningen research examines relations between agricultural innovation and social capital, regarded as one of the fundamental factors for sustainable human societies and communities.

Rijn (2014) investigates two types of development initiatives. The first is the implementation of agricultural research through the Integrated Agricultural Research for Development (IAR4D) approach. IAR4D was adopted by the Sub Saharan African Challenge Program (SSACP) and implemented in eight different countries. The second is the implementation of sustainable certification schemes through group-based experimental learning approaches. According to Rijn, social capital is associated with agricultural innovation. Second, development initiatives can influence social capital. Third, the existing level of social capital is associated with the success of development interventions. However, the effect was not necessarily positive and it depended greatly on the dimension of social capital. This argument is relevant to the Ethiopian context where relations between researchers and farmers need to be better understood. Researchers are naturally inclined to work with farmers, although they may have little experience or tools. Rijn's research concluded that bridging social capital which exists among people of like-mind is associated with the more extensive adoption of agricultural innovations, but negative associations are found between cognitive bonding social capital, which is supposed to be found among people with different background and interests. This difference in relations between social capital and adoption needs to be integrated in the up-scaling of participatory research in general and specifically in Ethiopia, where most of the policy implementation and local activities are carried out through the formation of groups in the locality and/or in organizations, known as the group approach. This is very common both in organizations and in agricultural development interventions.

Various understandings on participatory research by agricultural researchers in Ethiopia

Some researchers in Ethiopia strongly believe participation is an end in itself, the main goal of development. Amartya Sen (1999) proposed the idea of development and freedom, in which he explains that when human beings have freedom, they have more chance of developing as they each become agents for development. He also states that in the context of sustainability, we need a vision of humankind not as patients whose interests have to be looked after, but as agents who can effect change both individually and collectively. Robert Chambers has long been known as an advocate of 'putting the farmers first' in development and he initiated wide discussion on the failure of the conventional approach and made suggestions for alternative approaches (Chambers et al., 1989). He emphasized the importance of including the most vulnerable and excluded in the development process. Chambers also questioned the attitudes of researchers as embodied in the behavioural norms of professions, disciplines and bureaucracies in rural development. He has been recognized as a 'methodology focused activist' (Biekart and Gasper, 2013), which probably comes from his rather strong emphasis on methodologies. However, his standpoint was clearly stated in his critical statement on the Paris Declaration on development cooperation. Chambers extracted this statement from the declaration: 'To monitor indicators of effective performance from aid, donors and partners need the capacity to manage the mutual harmonization of programmes to assess, measure and report on the result.' Then he added: 'The words people, poor, power and relationship are nowhere to be found' (Biekart and Gasper, 2013). Chambers understands participation as a tool for empowerment and not for management. Therefore, self-mobilization is the best level of participation. Some authors have argued that the extent of participation could be advanced or could evolve from passive participation to self-mobilization with intermediate steps, including participation in information giving, participation by consultation, participation for material incentives, functional participation and interactive participation (Pretty, 1995: 173).

Another important belief about participatory research among Ethiopian researchers is related to agronomical appropriateness in the context of farmers. A book by Richards (1985) is a classic in this discourse, starting from the viewpoint of indigenous/farmers' knowledge for technology innovation and improvement. For Richards, one of the major reasons for farmers' participation in research is that farmers are the actors who have the knowledge for technology innovations. The FRG approach applies this fundamental insight, although it is not always explicitly mentioned. Biggs (2010) confessed to being disillusioned in the 1970s with quantitative macro-economic modelling as a means of investigating the impact of the Green Revolution. He also mentioned the more recent recognition by CGIAR that many technologies then developed on-station were inappropriate to the needs of poorer farmers operating under climatic diversity and uncertainty. On the other hand, as early as 1994, Bentley (1994) argued that participatory research attracts interest more because of dissatisfaction with the Green Revolution approach than because of the potential for success of collaborative work between farmers and researchers. He also suggested that participation works best in research agenda-setting and it is necessary for researchers to dedicate themselves to it for some time. Biggs (2010) analyses that there are simple dichotomies between the good aspects of participatory research and the bad aspects of non-participatory research in the 1990s, but this kind of reductionism is the object of critical consideration by promoters and followers of participatory research. It was another decade before this apparent paradox was overcome.

Farmers' Research in Practice edited by Veldhuizen et al., (1997) discusses many cases from the time of publication and includes recommendations for the improvement and transformation of research methodologies and institutionalization. Up to this point, many researchers and practitioners applied participatory research, but the outputs from such research did not meet their expectations. Participatory research within general agricultural and rural development was also discussed in many books during this period. Many institutes, mainly NGOs and think-tanks, published various guidebooks and project reports related to participatory research. They emphasized participatory methods and tools such as PRA and facilitation procedures, and some publications also explained the philosophy of participation (Pretty et al., 1995).

Before discussing different management strategies and the different stakeholders involved in participatory research, it is worth mentioning some research dealing with output-oriented discussion on agronomic findings. Hoffman et al., (2007) argue that it should be accepted that there are epistemological differences between farmers and researchers, and both should be allowed to work differently to create their respective comparative advantages. The particular proposal is that researchers' function in participatory research is to make the tacit knowledge of farmers explicit and feed it in to further formal research processes. This idea may be able to form the bridge between the thoughts of agronomists and the managers of research institutes.

A review of management mechanisms, including researchers and management staff in the research system, has been actively carried out in Ethiopia. In Europe, the key words are 'compromised participation', first expressed by Buhler et al., (2002). In their book *Science, Agriculture and Research: A Compromised Participation?*, they postulated that the intrinsic structure of agricultural research as a top-down activity in advanced nations (especially in the UK) made it difficult to change the research system and researchers' attitudes to a bottom-up one. They presented a clear understanding of the difference between participation in participatory research methods and participation in empowerment using the tools utilized in research activities. For the former, the method can be recognized as an efficient means of collecting data under complex conditions. For the latter, the empowerment aspect evolved partly as a response to the failure of extension systems to achieve adoption of technologies by farmers. The authors argued that the participatory approach was little more than a better method for technology delivery. In this sense, owners of the approaches did not change from researchers and government to supposed beneficiaries, in this case farmers. Another important aspect of management is a cost-benefit analysis of participatory research compared with conventional research. Although it is extremely difficult to compare these due to different scales and units of costs necessary for the two approaches, Neef (2010) showed from the experience of research projects in Thailand that the approach is cost-effective in blending scientific and local knowledge, scaling up micro-level data, and highlighting the constraints for farmers' technology adoption. This efficiency aspect was also supported by CGIAR centres during the 2000s, but mainly from the research institutes' point of view without seriously considering farmers' costs, including opportunity costs.

Efficiency and empowerment debates need more clarification of participatory research objectives in a given environment, owing to different requirements of different kinds and levels of stakeholder participation, although seeking efficiency and empowerment is not a mutually exclusive proposition (Aw-Hassan, 2008).

Owing to the diverse conditions in Sub-Saharan Africa, the necessity of a more systems-based approach has been proposed by Spielman et al., (2009), including an assessment of stakeholders' response to innovation and opportunities and constraints experienced by farmers, researchers and other stakeholders. This can be seen as a 'spinout' approach from participatory research, moving from a single stand approach to a more integrated management approach. The recent trend in research and debates on PAR has shifted more and more to this kind of systems-based approach, where words such as 'innovation', 'platform' and 'multi-stakeholder approach' are receiving attention in the global research arena (Leewis et al., 2014). These reports and conferences also lack concrete evidence for technical processes and innovations (for example, see Wettasinha et al., 2014).

This discussion is mainly for readers to understand the unique situation of Ethiopia in the process of institutionalizing participatory agricultural research when compared with global debates on the subject. When researchers in Ethiopia try to implement participatory research, they need to be aware of the power balance between stakeholders, especially between farmers, research management and administrations. In the next section, what sort of framework is required to understand the process of institutionalizing participatory research in Ethiopia will be discussed, based on the experiences of FRG projects over 10 years and the findings presented in previous chapters.

FRG experiences and key lessons

Through the activities of developing the FRG guidelines, conducting FRG approach training, supporting FRG-based research activities and developing the researchers' capacity of extension material development, the FRG I and II projects have had successes and failures in promoting the participatory research approach in the Ethiopian agricultural research system.

The FRG guidelines were developed to provide agricultural researchers with a practical guide to implement FRG-based research activities. The guidelines consist of straightforward steps with practical examples. FRG approach training was provided at six research organizations acting as training hubs and each trainee had to pass the three-step training programme. Through this process, a total of 600 researchers were trained. There were more than 80 FRG-based research activities, which were directly supported by the FRG I and II projects. The research activities tested the applicability of the FRG approach and their experiences were incorporated into the FRG guidelines. The development of extension materials by researchers was part of the process, as a research output in themselves. In order to facilitate the development of quality extension materials, researchers were provided with hands-on training. Some of the developed materials were tested with extension workers and farmers. The experience from this was that participatory research in general and the FRG approach in particular could deliver improved research outputs by integrating farmers' knowledge and researchers' scientific knowledge to achieve improved efficiency in terms of adoption and wider application. The approach also enabled farmers to participate in development as their right. The FRG approach has been extensively adopted, although to different degrees, by large numbers of researchers, some under development projects and others within normal research activities with government funding.

Despite its potential to improve the quality and efficiency of research delivery, there have been some challenges in realizing the potential of the FRG approach. The challenges fall in the areas of researchers' scientific capacity, their attitude towards farmers and farmers' knowledge, and organizational culture and working practices. Researchers' basic skills in designing research and data handling need to be improved, in addition to skills of communication with farmers and interpretation of farmers' knowledge, which are specific requirements of participatory research. Farmers' attitudes to researchers' 'supremacy' over farmers and other stakeholders is not easy to change, but its importance has been recognized by some researchers through field experiences, where farmers provided ideas and information crucial for successful research outputs. Through such experiences, the so-called 'unlearning' process where researchers do not insist on their superiority because of their academic career – has proved important, and 'knowledge' has been understood by researchers as a very individual matter. One typical example is described in Chapter 5 of this book (Laekemariam et al., 2016), which describes a process of change of research methodology by university researchers based on comments and opinions given by FRG farmers. The need for proper monitoring of field research activities and the follow-up actions of final research review and research report writing are not specific to but are critically important in FRG-based research, and guidance on these skills is not easy to get owing to the limited experience in participatory research among senior researchers. Sharing information and responsibilities within the research team was often missing and this made it difficult to complete research activities, especially where there is staff turnover, which is common in Ethiopia's agricultural research system. These challenges hindered the effective and efficient implementation of FRG-based research activities. It will be necessary to address these challenges strategically for future application of the approach.

Contribution of this book to enhancing research for development

As shown in previous chapters of this book, many researchers do not distinguish research activities and extension activities when conducting participatory research. As a result, many so-called participatory research activities do not follow the required procedures to draw scientifically convincing conclusions. On the other hand, such 'research reports' tend to emphasize the achievements in extension activities, such as the numbers of farmers trained, or the numbers of farmers who received seeds of improved varieties. Such researchers might believe that it is a shortcut to improving the livelihoods of farmers. However, it is something others, such as extension agents and NGOs working in rural development, can do – or at least do better than the researchers. For the researchers to contribute to the development of the agricultural sector in the country, it is more important to focus on 'research', fulfilling scientific requirements. Accumulating evidence in responding to farmers' needs eventually leads to the improvement of farmers' situation. Although it might seem to be a detour, it is actually required for sustainable development, and it is something researchers can do best.

At the institutional level, it is essential to accumulate information and foster information sharing. It is also important to evaluate the researchers' performance based on not only what the researcher has achieved, but also how well the researcher has addressed farmers' needs. It cannot be assumed, however, that by doing that, the researchers will be incentivized to respond to actual farmers' needs and the quality of research outputs will be secured. The same applies to training the researchers in general. It is of course important to present the ideal of how things should be in the training, but it is equally or more important to discuss how to address the reality. Besides, to assure the quality of the training activities to not only follow the theory but also to recognize concrete issues on the ground, it is experienced researchers' responsibility to monitor and mentor junior researchers. In this way, junior researchers will be evaluated not only on the amount of work they do but also on the quality of the process of research activities.

There is no shortcut in research activities to facilitate the development of farmers' livelihoods. However, by taking the necessary steps to ensure any research results add new information to previous research findings, researchers can eventually contribute to the development of rural livelihoods. The FRG II project in this aspect has contributed tremendously to assuring both the quality of the process of research and of research outputs with the FRG approach, and such achievements will be a foundation for forthcoming research to build on. It has been demonstrated through FRG activities during the projects that reinforcing the part played by research (scientific activities, communication with farmers, and facilitation among stakeholders) in participatory research does eventually improve the process of innovation. Thus, from the experiences of the project, the FRG approach could be said to be one of the options to realize a functional innovation system.

Contribution of this book in an academic context

Participatory agricultural research in Ethiopia needs to emphasize output-oriented research based on rights-based development because NARS is a part of the national administrative system based on national policy which emphasizes rights-based development, and because each research institute is evaluated based on its technical outputs and extension activities. The following three dimensions are a guide to understanding how researchers in Ethiopia consider participatory research within their institutional frameworks:

CONCLUSION: THE RESPONSIVENESS OF AGRICULTURAL RESEARCH SYSTEMS

1. participation as a human right,
2. participation for quality, as a means of integrating farmers' knowledge into research work as well as nurturing farmers' innovation capacity,
3. participation for quantity and scale, as a means of increasing the efficiency of research and technology adoption (Figure 13.1).

From the records of the projects described in this book, it can be concluded that participatory agricultural research through the FRG approach in Ethiopia has mainly been recognized as a methodology to improve the efficiency of research and extension relating to the problems faced by farmers in different agro-biological and socio-economic environments. Adoption of technologies is recognized as the single most important measurable indicator for the success of participatory research. The fact that actual activities in the field directly affect the behaviour and attitudes of stakeholders, including both researchers and farmers, underlines the weakness of the fashionable academic discussion of the 'innovation platform approach', in which the management aspects of participation are emphasized without much attention given to the technologies themselves, assuming that the research results, especially developed technologies, are appropriate and adoptable ones to farmers.

The quality of research outputs as scientific articles has also been strongly recognized by many researchers directly involved in FRG activities, including senior researchers in management positions. The attitude towards the scientific quality of participatory research has changed significantly from that of participatory research as an activity supporting extension work to one of

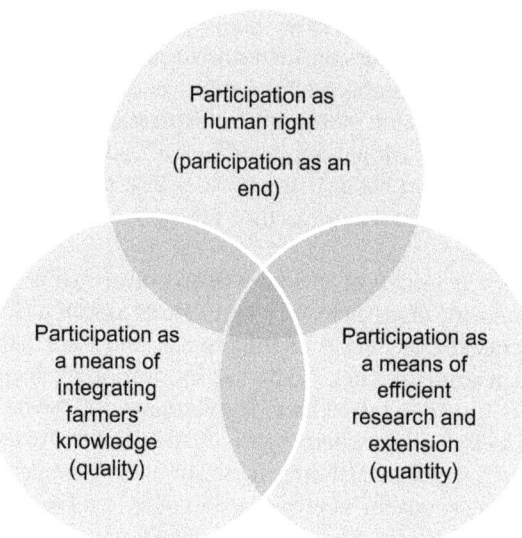

Figure 13.1 Three dimensions of participatory agricultural research

being an indispensable ingredient of the research-extension synergy. It is still the case, however, that the attitude of learning from farmers and taking on board the scientific background for participation has not been well recognized by some researchers. They tend to apply the guidelines as instructed, without modifications appropriate to the local situation. A typical example of this is the consideration of gender. In FRG research, it has been reported that a certain proportion of members were selected from female farmers, without any clear understanding of the roles of women for the particular technologies the research was examining. If the particular technology benefits only males within a particular social context, and/or the decision is made by males only, the simple inclusion of female farmers does not provide much benefit. The inclusion mechanisms for gender aspects, such as consultation, dialogue and demonstration, should be better recognized, even when the roles and functions of males and females are not clear in the target society.

The results of analysis described in Chapter 10 (Alemu and Shiratori, 2016) show the difficulties of applying participatory research in existing institutions, owing to a lack of capacity both in terms of individuals and organizations. Unfortunately, even university graduates did not have experience of note-and record-taking during field surveys and interviews. However, it became clear that continuous intervention by experts has changed the record-keeping and analyses done by FRG researchers and the importance of hands-on experience by each member of the research community has become clear with concrete examples. This was not reported in previous research articles.

Participation as a right in development has not been explicitly recognized by researchers, but it is observed that the political right of participation is a long-held government policy in Ethiopia and it is therefore not necessary to point out this aspect for particular activities. Many senior researchers expressed that the degree of participation was considered an important aspect of participatory research. Since the authors observed the shortage of training on participatory methods within the research system, these expressions by researchers can be recognized as an unintentional application of the development ideology of participation as an end in itself. Using a wider interpretation of the ideology of participation, participatory research in Ethiopia could be institutionalized more systematically.

Since the major objective of this book is to contextualize FRG approaches within the framework of participatory agricultural research and to share the concrete and detailed experiences of the projects and their research activities, the recent trend in academic circles to discuss innovation platforms and network approaches is not much discussed here. The contents of this book aim to fill the gap found in such discussions, where there is often little direct evidence of positive impacts from such research activities in relation to resource-poor farmers in Africa, or knowledge about the new transaction costs involved with engagement in innovation platforms[1]. In conclusion, the experience of the FRG projects has contributed to the discussion regarding the balance between the three different aspects of participatory research and their practical implementation.

Final words – a step forward

From our 10 years' experience of implementing and promoting FRG approaches in the Ethiopian NARS, we are confident that the FRG approach can contribute positively to the improvement of research and its practical application to the issues farmers face in their diverse agro-ecological and socio-economic environments. This is because of the intrinsic merits of FRG, which include allowing researchers to interact with farmers on specific needs, and their subsequent inclusion in the scientific research design. The approach also allows local knowledge to be considered in the research process, and the promoted group action and farmers' improved skills of experimentation in applying the approach ensure a continuous questioning of the status quo of farmers' livelihoods, often initiating other follow-up research and development efforts.

This in turn suggests the importance of continuous efforts to use FRG-based research on the ground, in order to build the capacity of the researchers belonging to the NARS and to institutionalize the approach more systematically. In addition, it will be important to incorporate the approach into formal higher education programmes as a research methodology. Further analysis on factors of success and difficulties encountered will continue to be necessary, to build the institutionalization process. This involves not just scaling up the model, but continuous day-by-day activities of the existing institutions and their members. Detailed descriptions of the processes, outputs, key challenges and opportunities outlined in this book will contribute to this process.

Note

1. For example, see the blog by Jim Sumberg at Future-Agriculture entitled 'Can "value chains" and "innovation platforms" boost African agriculture? Eleven reasons to be sceptical', <http://www.future-agricultures.org/blog/entry/can-value-chains-and-innovation-platforms-boost-african-agriculture-11-reasons-to-be-sceptical>.

References

Alemu, D. and Shiratori, K. (2016) 'The participatory approach and FRG: the institutionalization process within the Ethiopian agricultural research system', in Alemu, D., Nishikawa, Y., Shiratori, K. and Seo, T. (eds), *Farmer Research Groups Institutionalizing Participatory Agricultural Research in Ethiopia*, Rugby, UK: Practical Action Publishing <http://dx.doi.org/10.3362/9781780449005.010>

Almekinders, C., Proost, J. and Hagmann, J. (2009) 'Preparing scientists for society – A PhD training programme at Wageningen University, the Netherlands' in Almekinders, Beukema and Tromp, (eds.) *Research in Action – Theories and practices for innovation and social change*, Wageningen Academic Publishers, Wageningen.

Aw-Hassan, A. A. (2008) 'Strategies for out-scaling participatory research approaches for sustaining agricultural research impacts', *Development in Practice*, 18(94-5), pp. 564-575.

Bentley, J.W. (1994) 'Fact, fantasies, and failures of farmer participatory research', *Agriculture and Human Values*, 11(2-3), pp. 140-150.

Biekart, K. and Gasper, D. (2013) 'Interview with Robert *Chambers*', *Development and Changes*, 44(3), pp. 705-725.

Biggs, S. (2010) 'The lost 1990s? Personal reflections on a history of participatory technology development', *Development in Practice*, 18(4-5), pp. 489-505.

Blaakman, L. (2003), The art of facilitating participation: unlearning old habits and learning new ones', *PLA Notes*, 48, pp. 15-20.

Buhler, W., Morse, S., Arthur, E., Bolton, S. and Mann, J (2002) *Science Agriculture and Research: A Compromised Participation*, Routledge London.

Chambers, R., Pacey A. and Thrupp L.A. (eds.) (1989) *Farmer First: Farmer Innovation and Agricultural Research*, Rugby, Practical Action Publishing (Intermediate Technology Publications).

Blandon Cordoba, D. M. (2014) 'Participation, Politics and Technology: Agrarian development in post-neoliberal Bolivia', dissertation, Wageningen University.

Hagmann, J. and Almekinders, C. (2003) Developing 'soft skills' in higher education. *PLA Notes*, 48, pp. 21-25.

Hoffmann, V., Probst, K. and Christinck, A. (2007) 'Farmers and researchers: how can collaborative advantages be created in participatory research and technology development?' *Agriculture and Human Values*, 24(3), pp. 355-368.

Laekemariam, F., Gidago, G. and Taye, W. (2016) 'Lowering teff seeding rate using a seed spreader via the participatory approach in South Ethiopia', in Alemu, D., Nishikawa, Y., Shiratori, K. and Seo, T. (eds), *Farmer Research Groups Institutionalizing Participatory Agricultural Research in Ethiopia*, Rugby, UK: Practical Action Publishing <http://dx.doi.org/10.3362/9781780449005.005>

Leewis C, Schut M, Waters-Bayer A, Mur R, Atta-Krah K and Douthwaite B (2014) Capacity to innovate from a system CGIAR research program perspective. CGIAR Research Program on Aquatic Agricultural Systems. Program Brief: AAS-2014-29. Penang, Malaysia.

Li Jingsong, Lammerts van Buerenc E. T., Huangd K., Qind L., and Song Y. (2013) 'The Potential of Participatory Hybrid Breeding', *International Journal of Agricultural Sustainability*, 11(3), pp. 234–251.

Neef, A. (2010) 'Integrating participatory elements into conventional research projects: measuring costs and benefits', *Development in Practice*, 18(4-5), pp. 576-589.

Neef, A and Neubert, D (2011) 'Stakeholder participation in agricultural research projects: a conceptual framework for reflection and decision-making', *Agriculture and Human Values*, vol. 28, pp. 179-194.

Okali, C., Sumberg, J. and Farrington, J. (1994) *Farmer Participatory Research: Rhetoric and Reality*, Rugby, Practical Action Publishing (Intermediate Technology Publications).

Pimbert, M. (2004) *Institutionalizing participation and people-centered processes in natural resource management*, The International Institute for Environment and Development, London, and the Institute of Development Studies, Brighton.

Pimbert, M. (2012) 'Putting Farmers First', The International Institute for Environment and Development, London.

Pretty, J. (1995) *Regenerating Agriculture Policies and practices for sustainability and self-reliance*, London, Earthscan.

Pretty J., Guijt I., Scoones I. and Thompson J. (1995) *A trainer's guide for participatory learning and action*, The International Institute for Environment and Development, London.

Richards, P. (1985) *Indigenous Agricultural Revolution: The Ecology and Food Production in West Africa* London, Hutchinson & Co.

Rijn, F. C. van (2014) 'Social capital, agricultural innovation and the evaluation of agricultural development initiatives', Wageningen University.

Sellamna, N-E. (1999) *Relativism in agricultural research and development: is participation a post-modern concept?* London, Overseas Development Institute.

Sen, A. (1999) *Development as Freedom*, Oxford, Oxford University Press.

Shepherd, A. (1998) *Sustainable Rural development*, Basingstoke and London.

Spielman, D.J., Ekboir, J. and Davis, K. (2009) 'The art and science of innovation system inquiry: applications to Sub-Saharan African agriculture', *Technology in Society*, 31(4), pp. 399-405.

Veldhuizen, L., Waters-Bayer, A., Ramirez, R., Johnson, D. and Thompson, J. (eds.) (1997) *Farmers' research in practice: lessons from the field*, Rugby, Practical Action Publishing (Intermediate Technology Publications).

Swaans, K. (2014) *Study on impacts of farmer-led research supported by civil society organizations*, CGIAR Research Program on Aquatic Agricultural Systems, Working Paper: AAS-2014-40 Penang, Malaysia.

Index

Abule, E. et al. (2011) 81
action planning 18
active participation 20, 159
Adami Tulu Agricultural Research Centre *see* ATARC
Adesanwo, O.O. et al. (2009) 85
Adet ARC 135
AEEG (Agricultural Economics, Extension and Gender Research Directorates) 151
Africa Highland Initiative 2
Agricultural Growth Program, MoA 27
Ahmed and Eltegani (2012) 95
Alamata Agricultural Research Centre 178
Amhara region 89, 170, 172, 178
analysis of variance (ANOVA) 94, 110
Anwar, P. et al. (2011) 75, 76
Assosa Agricultural Research Centre 157
ATARC (Adami Tulu Agricultural Research Centre) 5, 26, 29, 120, 136, 152

'backstopping' 35, 156, 160, 162
Bahir Dar University 29, 135
Bako ARC 135
barley 105, 123
Benishangul Gumuz region 157, 169, 170, 172
Bentley, J.W. 190
Biggs, S. 190
Blaakman, L. 188
Blandon Cordoba, D.M. 188
BoA (Bureau of Agriculture) 167
Bolivia, participatory research in 188

Boran cattle 120
Boran x Jersey cattle 120, 122, 126, 128
Buhler, W. et al. (2002), *Science, Agriculture and Research: A Compromised Participation?* 191

capacity for innovation
  and education 188
  FREG approach 167, 173, 175
  FRG approach 5, 14, 15, 20, 21, 22, 46, 54, 66, 124, 143
  and institutions 196
  PPB and PVS approach 148
  support for 31
  see also CBID; Rural Capacity Building Project
Catley, A. et al. (2007) 137
CBID (Capacity Building in Irrigation Development) 37
Central Rift Valley (CRV) *see* common beans farming
cereal crops 49, 104, 105, 122, 175
  see also barley; maize; wheat
CGIAR 190, 191
Chambers, Robert 189, 190
chickpea production 178
Client-Oriented Research *see* COR
collective action 14, 15, 23, 186
commitment
  constancy of 114
  in FREG 170, 171
  of researchers 160–1
  as selection criterion 50, 91, 109
common beans (*Phaseoulus vulgaris*) farming, Central Rift Valley 47–66, 67, 149
  average seed yield 55–9

bean stem maggot 55
challenges to 65
farmer-based seed production 60, 62, 63, 64
features of 49
field visits 54, 55
group discussion 51
intercropping 60
moisture stress 55, 56
productivity 47
qualitative evaluation 56, 60–2
as security crop 47, 50
varieties 48–9, 51–4, 57–9, 61, 62
communication
and FREGs 180
and indigenous knowledge 161
information-sharing 15, 41, 65, 73, 186, 194
inter-group 54
lack of 39, 41, 65, 141, 142
and NARS 133
training hubs 32
Cool Season Food and Forage Legumes Project 2
COR (Client-Oriented Research) 4
cost sharing 14, 15, 65, 66, 72, 84
cropping calendars 26
CRV (Central Rift Valley) *see* common beans farming

DAE (Department of Agricultural Economics) 149
DAEFSR (Department of Agricultural Economics and Farming System Research) 13, 14, 149
daily calendar 26
dairy production, Melkassa 117–30
 artificial insemination 127
 crop production 122, 123
 crossbred cattle 118, 119, 120, 124, 127, 128–9
 data collection 119, 120
 environmental damage 127
 farmer demographic 119–20, 121–2

forage production 118, 127
gender and labour division 124–5, 128, 130
land allocation 123
livestock types 123
milk yield 118, 124, 125, 126
open grazing 118
record keeping 128
roles of partners 122
site selection 120, 121
smallholders 118
zero grazing 127
DAs (development agents) 122
 and formation of FRG 17, 18–19, 71, 73, 90, 108, 109, 152, 159
 FREGs 166–8, 171–5, 178, 179, 180
 and innovation 15
 movement of 65
 Debre Zeit Agricultural Research Centre 178
Degebassa, A. 95
Dembia district 90
Department of Agricultural Economics and Farming Systems Research *see* DAEFSR
Deutsche Gesellschaft für Internationale Zusammenarbeit (GiZ) 37
development agents *see* DAs
diagnostic surveying 21, 149
Dinka, H. 118
Dire Dawa BoA 175

economic feasibility 16, 81
EIAR (Ethiopian Institute of Agricultural Research)
 FRG implementation 2, 5, 14, 23, 46, 147, 148, 166
 FRG institutionalization 150, 151
 funding 36–7
 PVS and PPB 148, 149, 150
Enderta *woreda*, seed purification research 105–15
 data collection 109, 110

results 110–13, 114
selection of members 108–9
sites 105
team organization 109
environmental issues 46, 114, 127, 129, 130
Erer *woreda*, cereal crop farming 175
equal partnerships 15, 188
experience-sharing 19
extension workers *see* DAs (development agents)

FAO (Food and Agriculture Organization) 45, 46
Farmer Research and Extension Groups *see* FREGs
Farmer Research Groups *see* FRGs
farmers' participatory research *see* FPR
farming system compatibility 16, 99
Farming Systems Approach *see* FSA
Farming Systems Research *see* FSR
FDG (Farmers' Development Group) 27
Fedis ARC 135
fertilizer rates 73, 75, 85, 138, 157
FGDs (Focus Group Discussions) 166, 167
field activities 18, 19, 53
fish processing, Lake Tana 89–100
  data analysis 94
  drying duration 91–3, 94, 95
  effect of technology on price 97, 98, 99
  evaluation 96
  exports 89, 90
  'field days' 100
  FRG participation 93–4, 99, 100
  livelihood 89
  microbial analysis 94, 95, 97, 98
  moisture content analysis 93
  organoleptic testing 93, 95
  perceptions of final product 99
  perceptions of technology 98, 99
  selection of members 90, 91
  shelf life 97, 98
  smoking 91
  total plate count 95
  traditional drying method 92
Focus Group Discussions *see* FGDs
Fogera district 90, 99, 135
Food and Agriculture Organization *see* FAO
food insecurity 2
forage seed production 159
FPR (farmers' participatory research) 149
FREGs (Farmer Research and Extension Groups) 20, 21–2, 152
  characteristics of 169–70
  cost-benefit analysis 166, 173–7, 180
  establishment of 167
  exit strategy 178, 179, 180
  focal persons 174
  funding 176
  member farmers 170–3, 177
  perceived benefits of membership 173–4, 177
  performance of 179, 180
  promotion of 167, 169
  sustainability 177, 178, 179, 180
  technical support and supervision 171, 174–5
FRGs (Farmer Research Groups)
  adoption of technology 137, 138, 139, 140, 141, 142
  assessment of farmers' needs 16
  challenges 137, 139, 140, 142, 155–63
  as collaborative 150
  diversity of membership 15
  farmers' perceptions of 133–43
  formation 14, 17–18
  framework 14, 15, 18, 20, 23, 28
  gender guidelines 25–41, 192–3
  importance of communication 142
  innovation

institutionalization of 147, 148, 150–3
as multidisciplinary 16
linkage
organizational culture 161
peer reviews 159
planning activity 15, 16
process 15, 16, 17
research ethics 159–60, 161
research overview 13–23
researchers' technical capacity 156, 157–9
review, monitoring and evaluation 160
scaling up 165–80
scientific recognition of 150, 151
technological options 16, 17
text mining 136, 137
training 28–32, 150–3
FRG I
development of guidelines 26
implementation of 5, 14, 25, 46
institutional arrangements 20, 23
FRG II 133, 155, 161
implementation of 5, 14, 25
institutional arrangements 20, 23
procedures research 162
research 32–40, 134, 194
research ethics 160
training 28–30, 151
use of guidelines 26, 27, 28
FSA (Farming Systems Approach) 166
FSR (Farming Systems Research) 4, 13, 21–2, 147, 148–9, 165
Fufa, H. et al. (2001) 76

gender analysis workshops 26
gender roles
dairy production 124–5, 128, 129, 130
female farmers 26, 52, 50, 59, 119, 121, 125, 128, 161, 169, 196
FREG membership 169
FRG membership 150

household roles 122
mainstreaming guideline 169
GiZ see Deutsche Gesellschaft für Internationale Zusammenarbeit
goat farming 175
Gobeze, L. et al. (2007) 75, 76
Green Revolution approach 190

Haramaya University 166, 175
Hawassa University 29, 31
Hirna Irrigation scheme 37
Hoffman, V. et al. (2007) 191
human right, participation as 195

IAR (Institute of Agricultural Research) (now EIAR) 4, 13, 148
IAR4D (Integrated Agricultural Research for Development) 189
IDRC (International Development Research Centre) Canada 2
IIED (International Institute for Environment and Development) 187
indigenous knowledge 133, 158, 161, 165, 190
information-sharing see communication
'innovation platform approach' 195, 196
insect pests 54, 66, 70, 73, 135, 138, 158
interactive participation 20, 21, 23
International Commission on Microbiological Specifications for Foods 95

JICA (Japan International Cooperation Agency) 2, 5, 14, 20, 36–7, 108
Jones, A.L. 136

KIIs (Key Informant Interviews) 166, 167
Kulumsa Agricultural Research Centre 105

Lake Tana *see* fish processing
lentil production 178
Li Jingsong et al. (2013) 187
literacy 128
livestock 117, 119, 123–5, 127, 159, 170 *see also* goat farming; oxen

M&E *see* monitoring and evaluation
maize 50, 60, 93, 123, 127, 158, 177
MARC (Melkassa Agricultural Research Centre) 5, 20, 26, 29, 46, 121
marketing
　collective 15
　common beans 47, 56, 64, 65
　dairy products 118
　fish processing 90
　seed 148, 179
May-Tsuberi ARC 136
Mekelle University 29, 31, 104, 108, 135
Melkassa *see* dairy production
Melkassa Agricultural Research Centre *see* MARC
Minjar Shenkora *woreda* 178
MoA (Ministry of Agriculture) 27, 32
Mohammed (2011) 90
monitoring and evaluation (M&E)
　importance of 19, 35, 156, 160, 193
　interactive participation 20, 167
　support for 37, 38
Moran, J. 118
multidisciplinary teams 15, 109, 149, 156
Mulu et al. (1994–5) 70, 79

NARS (national agricultural research system)
　different approaches 148
　and FRG 5, 14, 25, 37, 133, 147, 166, 186, 197
　institutionalization of FRG 150–3
　rights-based 194
Nathareth (now Melkassa) 2

national lowland pulse research programme 47
Naturland, V. 70
Neef, A. 191
Neef, A. and Neubert, D. 187
NGOs (non-governmental organizations) 2, 188
Nishikawa, Y. 81
NVRC (National Variety Release Committee) 150

OARI (Oromia Agricultural Research Institute) 4, 5, 14, 20, 46
Office of Agriculture and Rural Development 104, 108
Ofla *woreda* 178
Okali, C. et al. (1994) 186
organoleptic testing 93, 95
Oromia Region 14, 20, 46
output-oriented evaluation systems 9
Owuor, B.O. et al. (2001) 70
oxen 73, 123, 126

PAPRG (Pastoral Agro-Pastoralist Research Group), guidelines 27
Paris Declaration 190
participatory approach 9, 10
　debates 186–9
　history of 1, 2
　and human rights 195–6
　features of 21–2
　training 188
passive participation 20
Pastoral Agro-Pastoralist Research Group *see* PAPRG
Pastoral Community Development Project 27
Pimbert, M. 187
PPB (Participatory Plant Breeding) 14, 148, 149, 150, 187
PRA (participatory rural appraisal) 26
PTD (Participatory Technology Development) 20, 21–2, 166

PVS (Participatory Variety Selection) 148, 149, 150

QSPP (Quality Seed Promotion Project) 36

RARIs (regional agricultural research institutes) 5, 148, 150–1
Rathore, P.S. 70
RCBP (Rural Capacity Building Project) 166
Regional Capacity Building Project 152
research implementation 32–40
  challenges to 37, 38–40
  communication and 41
  data collection 38, 39, 40
  funding 36, 37
  identification of priorities 33, 34
  information-sharing 41
  IT training 40
  monitoring and evaluation 35, 37
  multidisciplinary teams 39
  need for incentives 41
  proposals 34, 35
  research reports 35, 36
  review meetings 39, 40
  support 33, 36–7
research capacity 17
research-extension process 14, 150–1
research, targeted 134–6, 141
resource mapping 26
rice, fertilizer research 157
Richards, P. 190
Rijn, F.C. van 189
risks, expected 16
RREP (Rural Resilience Enhancement Project) 37
Rural Capacity Building Project *see* RCBP

Sato, I. 136
self-mobilization 187, 190
self-reflection 188
Sekota ARC 135

Sen, Amartya 189
social capital 189
social feasibility 16
Software Package for Social Science (SPSS) 94
Sogido-Saraweyba Irrigation scheme 37
solar tents 91, 92, 93–5, 97, 98, 99–100
Spielman, D.J. et al. (2009) 192
spousal membership 26
SSACP (Sub Saharan African Challenge Program) 189
staff turnover, training hubs 31, 32, 152
stakeholders, identification of 18
Stallknecht, G.F. et al. (1993) 81
systems-based approach 192

technology, adoption of
  challenges to 20
  collective action 15
  feasibility 16, 81
  impact on price 97
  perceptions of 98, 99, 137, 142, 177
  quantification 137, 139–41
  technical support 174
teff (*Eragrostis tef*) 69–85, 72
  agronomic data 73, 74
  continuous cultivation 85
  cost to benefit ratio 84
  crop development 73, 75–7
  data analysis 75
  design of participation 72, 73
  economic analysis 75, 83
  field evaluation results 74, 79, 80–4
  FPR 149
  harvest index 79
  multidisciplinary approach 71
  preparation 72
  seed spreaders 70, 71, 72
  seeding rates 77
  study area 71
  weed infestation levels 85
  yield 76, 77, 78–9, 82, 83

template forms 26
Tessema, A. et al. (2008) 90, 94, 95
Tigray region *see* wheat
training hubs 31, 32, 150, 151, 152, 202

'unlearning' 188, 193

Veldhuizen, L. et al (1997), eds., *Farmers' Research in Practice* 190

Wageningen University 188
WARC (Werer Agricultural Research Center) 29
weed control 69, 73, 85, 138
Werer ARC 136
wheat (*Triticum aestivum*) 103–15
  data collection 109, 110
  design of research 105–6
  evaluation 113

FREG 178
planting methods 106, 108
purified seed 103–4
salt solution purification 106–8
seed quality analysis 108, 112, 113
selection of members 108–9
socio-economic analysis 113, 114
trial sites 105
yield 103, 106, 110–13
Wolaita Sodo University 71, 135, 136
Wolaita Zone 48, 49 *see also* teff
Woreda Office of Agriculture and Rural Development 104
World Bank 27
Wukro College of Agriculture 31

Yenesew, Y. 76
Yimer, A. 90, 94, 95

www.ingramcontent.com/pod-product-compliance
Ingram Content Group UK Ltd.
Pitfield, Milton Keynes, MK11 3LW, UK
UKHW021826140426
5217IPUK00004B/108